The Art of Safety Auditing

Auditing
A Tutorial for Regulators

T0341219

The Art of Safety Auditing
A Tutorial for Regulators

Sasho Andonov

CRC Press
Taylor & Francis Group
Boca Raton London New York

CRC Press is an imprint of the
Taylor & Francis Group, an **informa** business

CRC Press
Taylor & Francis Group
6000 Broken Sound Parkway NW, Suite 300
Boca Raton, FL 33487-2742

© 2020 by Taylor & Francis Group, LLC
CRC Press is an imprint of Taylor & Francis Group, an Informa business

No claim to original U.S. Government works

Printed on acid-free paper

International Standard Book Number-13: 978-0-367-35108-3 (Paperback)
International Standard Book Number-13: 978-0-367-35761-0 (Hardback)
International Standard Book Number-13: 978-0-429-32979-1 (eBook)

Visit the Taylor & Francis Web site at
http://www.taylorandfrancis.com

and the CRC Press Web site at
http://www.crcpress.com

Contents

Acronyms and Abbreviations

ADP	Automatic Data Processing
AIS	Aeronautical Information Services
AMC	Acceptable Means of Compliance or Alternative Means of Compliance
ANSP	Air Navigation Service Provider
ATC	Air Traffic Control
ATCo	Air Traffic Controller
BP	British Petroleum
CAA	Civil Aviation Agency
CLs	Check Lists
CNS	Communication, Navigation, Surveillance
COM	Communication
DFS	Deutsche Flugsicherung
DGCA	Directorate General of Civil Aviation
DME	Distance Measuring Equipment
EASA	European Aviation Safety Agency
FAA	Federal Aviation Administration
FMS	Financial Management System
GCAA	General Civil Aviation Authority
GP	Glide Path
HF	Human Factors
HSE	Health, Safety, and Environment
IAEA	International Atomic Energy Agency
IANS	Institute of Air Navigation Services
IATA	International Air Transport Association
ICAO	International Civil Aviation Organization
IT	Information Technology
ISO	International Organization for Standardization
LLZ	Localizer
LMS	Logistic Management System
MET	Meteorological Services (in aviation)
MRO	Maintenance, Repair, and Overhaul (Maintenance and Repair Organization)
MTBF	Mean Time Between Faults (Failure)
MTC	Military Technological College
MTTR	Mean Time To Repair
NAVAIDs	Navigational Aids
NTSB	National Transportation Safety Board
QMS	Quality Management System
PMS	Production Management System

RDL	Radiation Dose Limit
RPN	Risk Priority Number
SAR	Search and Rescue
SeMS	Security Management System
SMS	Safety Management System
SSP	State Safety Program
SUV	Sport Utility Vehicle
USOAP	Universal Safety Oversight Audit Program
VHF	Very High Frequency
VOR	VHF Omnidirectional Range

Definitions of the Terms Used

Term	Definition
Risky Industry	These are the industries which can produce damage to, or deaths of, humans, assets, and the environment (on a large scale!) by their production process, by their products, by the processes of providing services, or by services provided. Such industries are nuclear, chemical, pharmaceutical, aviation, food industry, medical, etc.
Regulation	This is a sum of laws, rules, standards, or procedures which are accepted by a particular authority in the State and/or on the international level.
Regulator	This is a body established by the State (as part of the State Administration) which is tasked with, and responsible for, passing and overseeing the fulfilment of regulations in a particular area in the State.
Audit	One of the tools used to oversee the Management Systems (quality, safety, environmental, financial, etc.) in an industry. It is a process of overseeing (checking) the performance of various companies in the industry to check have they implemented, monitored, controlled, and maintained any Management System.
Auditor	Person employed by the Regulator. This person is educated, trained, and skilled in providing regulation and auditing the activities of the companies in the areas of responsibility of that Regulator.
Auditee	Person or entity (company) which is subject to auditing.
Audit Team	This is a team of auditors established to do auditing in the company because the audit is a complex job and one auditor cannot do it alone.
Incident	This is an event which produces harm to humans (animals, nature, etc.) and damage to assets.
Accident	This is an event which produces deaths for humans (animals, nature, etc.) or total damage to assets (they cannot be repaired and cannot be used anymore!).
Post-holder	This is a position in the company (in a Risky Industry!) which can be given to the person who is subject of Approval from the Regulator.
Hazard	It is a situation which has the capability of producing harm or death to people and damage or destruction of the assets.
Risk	When the hazard is quantified regarding the severity (how strong the consequences are, if it happens) and frequency (how often it can happen), we are speaking about Risk.
Risk Management	It is the sum of processes for Hazard Identification, quantification of hazards to obtain the Risks, risk elimination and/or risk mitigation, preventive and/or corrective actions, sharing of safety-related information, etc.
Fault	It is a term used to express a problem with the equipment which causes equipment to stop functioning, or the equipment to stop functioning within specified tolerances for normal functioning. Fault is connected only with equipment!

(Continued)

Term	Definition
Failure	It is a term used to express problem with an operation which is not conducted at all or is not within specified tolerances. Failure is connected only with operations. It is not always the case, but the faults of equipment can produce failures.
Preventive Actions	These are actions executed by the company with the intention to eliminate or mitigate the risk which was registered, and a particular analysis (done in advance!) showed that it can materialize if not treated. Simply put, these are actions which are done in advance to eliminate or mitigate the risks.
Corrective Actions	These are two types of actions: (a) Actions undertaken to fix a particular problem in the company which can possibly be the cause of incident and/or accident in the future or to fix non-compliance in the Management System. (b) Actions undertaken when an incident or accident happened and the consequences must be eliminated or mitigated.
Verification	This is the process of checking, testing, or approving some product, service, and/or Management System which specifications are requested by some international standard or by some regulatory requirements.
Validation	This is the process of checking, testing, or approving some product, service, and/or Management System for which specifications are not requested by some international standard or by some regulatory requirements, but are actually the product of a design by the company itself.
Accreditation	This is a formal process of recognition that a company or an individual is capable of doing particular job in accordance with standardized specifications (standards!).
Certification	This is a formal process of recognition that a company or an individual has achieved something which can be expressed by a particular level of excellence, education, implementation, or status.
Attestation	This is a formal process of recognition that a product is satisfying the specification required by particular standard in that area.
Approval	This is a formal process for approving a particular company or individual to do their job.
Mistake	A mistake is usually an outcome from a bad human choice. It is connected to free Human choice, usually when the lack of proper information, lack of understanding, or lack of knowledge is present.
Error	Error can be made by humans and equipment and is connected with the malfunction of something which is assumed to be OK (a code in computer software, an operation (human and equipment), a rule (human), etc.).
Safety Event	This is any event which, if it happens, will increase the belief that the safety performance of the company experiences problems.
Compliance	The level of accordance of the performance of the company with the regulation in that area.
"Personality"	This is a term which applies to the Regulator and to companies. The "Personality" of the Regulator can be defined as a combination of attitude, style, and character in doing oversight activities. The "Personality" of the company can be defined as a combination of attitude, style, and character of doing business.

Term	Definition
Effectiveness	It can be defined as the level of success of a particular process, operation, activity, task, etc. Effectiveness is achieved if the particular process (operation, activity, task, etc.) is successfully finished or it can be stated that the process (operation, activity, task, etc.) was successful to a high level, for example: The process was 93% successful.
Efficiency	This is an economic category and it can be roughly defined as the ratio of resources and production in the company. The more production and fewer resources, the better the company is (more profit is gained!).
Objective Evidence	For the purpose of auditing, it can be defined as data (in the form of records, written statements or commands, measurements, reports, etc.) presented to the auditor which will prove that the company is doing (or not doing!) what is required by regulation.
Scope of Audit	This can be defined as a measure for quantification and quality of details which will be considered during the audit. It depends from the type of audit.
Briefing	This is a meeting where short information is provided or explained to the employees regarding some issue, some activity, or some change.
Change	This is a change of the operation, activity, procedure, equipment, software, facility, employee, etc. for a new one.
Modification	This is also a change, but it affects just a part of the operation, activity, procedure, equipment, software code/program, facility, employee, etc. that is changed with the new one or to a new setting.
Warning	Signal (light, sound, etc.) which is triggered when the value of something dangerous (temperature, radioactivity, concentration of gasses, etc.) is coming close to the value when the control over the process can be lost, or the process (if continued) will reach values to damage humans, assets, or the environment.
Alarm	Signal (light, sound, etc.) which is triggered when the value of something dangerous (temperature, radioactivity, concentration of gasses, etc.) is exceeding the value when control over the process can be lost, and the process (if continued with these values) will damage humans, assets, or the environment.
Measurand	For the purposes of this book, the measurand is a value (temperature, radioactivity, concentration, etc.) which is measured to provide information regarding the normal performance of the processes (operations, activities, etc.) in Risky Industries.

Preface

I am shocked by the fact that in the engineering community in industry, there is a huge lack of proper understanding about quality! I am also shocked by the fact that in the engineering community in Risky Industries, there is huge lack of proper understanding about safety!

I started my "journey" in the area of Quality and Safety by command: One day in the summer of 2005, my boss came into my office and told me: There is a requirement from EUROCONTROL* to implement a Safety Management System (SMS) in our company and I choose you to undertake the responsibility to implement it in the Technical Department. At that time, I was working in the Planning and Development Department of the Civil Aviation Agency in the Republic of Macedonia, and I was familiar with the newest developments in aviation regarding improvements of safety. I was aware that this was a big movement forward for aviation subjects, but I wondered why we were implementing an SMS when a Quality Management System (QMS) can provide excellence in our operations.

The "journey" into the safety area which started that summer in 2005, I am still travelling today! Of course, (later) I realized why we need Safety Management and why Quality Management alone cannot provide better safety in all Risky Industries.†

Later, in my professional life, I had a few positions which were connected with implementations, maintenance, and auditing of the QMS and the SMS, and I do believe that I have reached a level of knowledge, skills, experience, and attitude in areas of Quality and Safety which make me satisfied. The point is that by increasing my knowledge and experience, I noticed that most of the people (in general) involved in industry do not understand what Quality and Safety are. I started to provoke discussions on Safety and Quality forums on the internet with the intention of improving understanding, but it was also a chance for me to gather new knowledge and new experience through these discussions.

In 2014, I started to work for the Military Technological College (MTC) in Muscat (Oman) as Senior Instructor EASA.‡ Our Aeronautical Department was approved as an EASA Part 147 Training Organization and we were the subject of EASA Regular Audits each year. Having in mind that by that time, I had already 12 years' extensive experience in the Quality and Safety areas,

---

* EUROCONTROL is the European Organization for the Safety of Air Navigation, with its headquarters in Brussels, Belgium.
† Risky Industries are aviation, transport, nuclear, chemical, petroleum, pharmaceutical, medical, etc. All industries which can produce extensive damage or deaths of Humans, assets, and the environment are classified as Risky Industries.
‡ EASA is the European Aviation Safety Agency.

I could not just run away from my "safety attitude." As an EASA Part 147 approved organization, we had to implement a QMS and an SMS, but in my humble opinion, these systems were implemented in a very wrong way. I tried to explain to my Head of Department that things were wrong, but it just brought me problems which made me quit that job later. I thought that maybe the EASA auditors would point to elementary deficiencies of our QMS and SMS, but that was another shock for me: The two auditors, during two different audits, did not registered any non-compliance!

I was very, very, very disappointed by EASA.

In 2015, I received an invitation to attend the Conference of Air Transport Operations (CATO 2015) and I submitted a paper with fresh examples of significant misunderstanding of Quality and Safety in airlines and MROs. After the presentation of the paper, a young PhD student from Deutsches Zentrum für Luft- und Raumfahrt (DLR) from Braunschweig sat next to me during lunch and said: I am shocked by your presentation. I asked him did he have some arguments against my presentation, so we could discuss that and he responded: No, I am shocked by the fact that whatever you said makes sense.

If the EASA, as one of the well-recognized and respected organizations regarding aviation safety in the world, cannot provide proper training for their auditors, then there is huge gap in aviation safety. And this is intention of this book: To fill a gap in regulatory auditing, by providing knowledge and advice to the Regulators (and their employed auditors!) of Safety Management Systems to do their job as it needs to be done!

Am I qualified to write this book?

Let's see my "credentials:"

- I have been working in the area of Quality and Safety in industry for the last 15 years.
- I have passed the exam for the SAF AUDIT course in EUROCONTROL IANS.
- I have passed exams for ICAO* USOAP† CMA Phase 1 and ICAO USOAP CMA ANS courses.
- I am an Internal Auditor of ISO/TS 16949 and a Lead Auditor of ISO 14001.
- I have worked as a Safety Auditor/Advisor in CAA of Macedonia.
- I have worked as a QMS Manager in a Hi-Tech Corporation (Skopje, Macedonia) where I implemented ISO/TS 16949 and AS (EN) 9100 and integrated them with the already implemented ISO 9001 (which was "reshaped" before integration!).

---

* ICAO is the acronym for the International Civil Aviation Organization (the specialized UN agency for aviation).
† USAOP stands for the Universal Safety Oversight Audit Program of ICAO.

- I have "survived" two external audits in the Hi-Tech Corporation (by auditors from Denmark and Serbia) for the Management Systems which I implemented in the Hi-Tech Corporation and there were no findings of non-compliance.
- I have worked as QA & Safety Manager (Post-holder) in Travel Air (airline from Madang, Papua-New Guinea) where, in few months, Copy-Pasted QMS and SMS from Quantas (Australia) were adapted to the reality of Travel Air.
- I have worked as a CNS Expert for ICAO in New Delhi (India) where I was helping DGCA of India to "establish safety oversight capability" in the area of CNS.
- I have held two courses on Modern Audit Techniques in Indian Aviation Academy in New Delhi.
- I have provided two on-the-job training sessions for ten of DGCA's CNS inspectors in India.
- I have already published two books in the USA regarding Quality and Safety.

So, yes: I do believe that I am competent for the job!

I hope the book will provide enough good material, and it will help (by following my recommendations inside) to provide excellent audits which will satisfy Regulators (auditors) and the companies (auditees).

Sasho Andonov

Author

Sasho Andonov is a Graduated Engineer in Electronics and Telecommunications and has earned a Master's Degree in Metrology and Quality Management at Ss. Cyril and Methodius University in Skopje, North Macedonia. Sasho has 24 years' experience in aviation, especially in the area of ATM/CNS and Quality and Safety Management, and 14 years' experience in standardization and accreditation. He is a member of the technical board in the electro-technics, information technology, and telecommunications area of the Institute of Standardization of Republic of Macedonia and a member of the Sectors Committee for electrics, electronics, and electrical machines of the Institute of Accreditation of Republic of Macedonia. Sasho has contributed to 13 international conferences and symposiums, with his papers mostly in the areas of satellite navigation, calibration and safety, and quality management. He has already published two books: *Quality-I is Safety-II: The Integration of Two Management Systems* and *Bowtie Methodology: A Guide for Practitioners.*

1

Introduction

1.1 Who Can Benefit from This Book?

This book is dedicated primarily for the auditors of Regulators in Risky Industries. The intention of this book is to explain to these auditors what is the audit's purpose and how the audit must be conducted to provide a Win-Win situation. By a "Win-Win situation," I mean: The Regulator is doing its job and the regulated company is happy with the overall performance!

Whatever you think about auditing, keep in mind that it is a complex activity, and to get real benefit from the audit, it needs to be conducted in the way which I define as a way of producing Art! Similar to the artist who strives to produce a masterpiece, the auditor must first become enough mature in knowledge, skills, and experience. After that, there are activities of preparation and there are activities of conducting the audit. As the particular dedication, attitude, skills, knowledge and experience are needed from the master artist, all these attributes are needed also from the auditor. As the artist puts a lot of effort into creating his masterpiece, so the auditor must put a lot of effort into providing a good audit.

Although it is a book for Safety Regulators, it applies for all types of audits done in the general meaning of the term "industry." Of course, any other industry (different from Risky Industries) has its own specifics which have to be taken into account during consideration of recommendations inside this book. Proper understanding of everything mentioned in this book will provide enough confidence to use this book in any other industries.

The main point with Risky Industries is that they are very well-regulated, so if there are rules on what to ask during the audit, it is easy to conduct the audit. Satisfying regulatory requirements for stronger regulation (as it is in Risky Industries) will help companies from other industries to deal with less strong regulation, as it is in their areas. And that is the reason that this book is a good tutorial for all types of auditors of any management system in all different parts of the industries.

I will try to explain in this book all the aspects of audits in a narrative way. Anyway, I will present some axioms (in bold and italic in the text!) which must not be violated! I will use my knowledge and experience gathered through my 20 years of professional life in aviation and industry, especially in the area of quality and safety management from the regulatory, service, and/or manufacturing point of view.

The book is dedicated primarily to Safety Regulator employees. How they establish their attitude in the particular Risky Industry is the way the regulated companies will behave. If the Safety Regulators earn the respect of the regulated companies, then they can be successful in overseeing the safety performance of the companies. The highest level of regulation behavior in the State Safety Regulators must provide an atmosphere of dedication, knowledge, skills, and attitude which, naturally, will be transferred to the regulated companies. In general, the Regulator's behavior overall can produce a lot of benefits or a lot of damage.

The book is written in a general style. There are many things which are explained inside and all this is done having in mind good practice in conducting audits. The reader can decide in many areas what to do. I tried to produce a holistic book about regulation and audits and I put there a lot of things (mostly explanations) why these things need to be done. As I mentioned in the Preface, I have seen a lot of misunderstandings regarding audits, and I tried to clarify in this book what, and why, they are wrong. From another side, I tried to emphasize good things which must be maintained to provide a good audit.

This is a book which needs to be read more than once. Reading just once will not make you a good auditor, so this is more a manual than a book.

This book can also bring benefit to the companies which are subject of safety regulation, especially for their managers. Providing safety is team work and the necessary partnership between Regulators and companies is of utmost importance. Without this partnership, where everybody is aware of their responsibilities and duties, any type of safety can hardly be as effective and efficient as it has to be.

1.2 General Explanation Regarding the Book

What is an audit?

If you go on the internet and write "what is an audit," you will have approximately a few hundred thousand hits. There are different definitions, some of them general and some of them sticking to a particular area (financial, quality, internal, etc.). For the purpose of this book, I will define the safety audit as a process of examination (checking) of various companies in Risky Industries to see if they have implemented, monitored, controlled, and maintained a Safety Management System.

The process of auditing applies to the documentation and to the real situation in the company, so it must be done off-site (documentation) and on-site (implementation, monitoring, control, and maintenance). Whatever you are thinking about audits, it is important to understand that auditors check the capability of the company to maintain control over production processes without endangering humans, assets, and the environment with its products or with the services offered.

The process of auditing, especially in Risky Industries is a complex activity. It comes from a more complex area called Regulatory Oversight. Audit is just one part of the oversight activities and it bears the same complexity as other oversight tools. Having in mind that auditing is a complex activity, there is a need for an Audit Team which will do the audit. The Audit Team consists of a Team Leader* and Team Members (auditors).

So, this book is also complex in its constitution. I will speak about different things and you must understand that all these different things have different meanings when they are alone, and when they are just part of the whole process of auditing. Please keep in mind that there is a need for all these parts to be synthesized in one whole activity called auditing.

There is another aspect of this book which I would like to emphasize. This is in regard to the terms used today in the industry and regulation: There is complete chaos and for the same thing, there are a few names which sometimes could be pretty much contradictory. So, in the next chapter, I will try to explain these "ambiguities" regarding particular terms and I hope it will be accepted in the regulatory and manufacturing community.

I will try to express the facts and my opinions and eventually, you will decide by yourself how it will change your understanding of audit activities. Anyway, this book is written from the practical point of view and it is based on my, and others', practical experiences. Please, keep that in mind when you read the book!

This is a place where I would like to give some general explanation of the terms used in this book. I find it very important because the book is written to be international, so different States, organizations, and/or Risky Industries will use different terms.

As is mentioned in the Definition chapter, by the term Risky Industries, I will understand the industries like aviation, transport, nuclear, chemical, petroleum, pharmaceutical, medical, etc. In general, these are industries which can produce harm and/or damage to the assets and environment and harm and/or death of humans. These industries are very well regulated, and there is regulatory requirement for each subject belonging to them, to establish, implement, maintain, and continually improve the Safety Management System (SMS).

I will use the word "Regulator" for the State entities which are established by each country as a part of the State administration with the task of

---

* Somewhere in the literature, you can find the expression Lead Auditor for the Team Leader. There is nothing wrong with that, but I prefer to use the expression Team Leader for the person who is in charge of Audit Team.

overseeing the subjects belonging to Risky Industries. It can be any type of agency, directorate, department in the ministry, or whatever it is named by particular State.

I will use word "State" with capital first letter with the meaning of state (country), such as USA, Germany, United Kingdom, etc. This meaning is different to the meaning of the word "state" as a situation, position (state of being OFF or ON or on standby for Equipment, or being healthy or ill with Humans), or verbal activity.

An auditor is a person who is doing the audit and could be employed by the Regulator, or he or she can be employed by a company which provides auditing services for the Regulator in a particular State in accordance with a signed contract. An auditee is a company (department, unit, or person) being audited.

I will use word "counterpart" for a company's employee who joins the auditor during the audit. During the audit, each auditor must be accompanied by such a person who will help the auditor. This person will provide valuable information for the auditor, and in addition, he will provide access to the company's premises and restricted areas during audits.

I will use word "equipment" with the meaning of hardware, software, and aggregation of the hardware and software (systems!) for the machines, tools, instrumentation, and whatever else is used in the company for manufacturing products or offering services. This is important to understand, because there are rules on how the hardware of the Equipment must be manufactured in Risky Industries, and the companies mostly follow these rules. But there are also rules on how the software should be produced and certified, and these rules are not always followed.

I will repeat a few times the same things in different places in the book. I found the expression *"Repetitio est mater studiorum"* (translation from Latin: Repeating is mother of learning) very useful during my "teaching sessions" with students, technicians, and engineers in my professional career, but the real reason for repeating them is that mostly these things have different contexts in different areas.

1.3 Axioms

Axioms are statements which come from mathematics. They are used there to provide statements which are accepted as TRUE and as such, they do not need to be proved. So, in this book, I am using axioms as statements regarding the audit rules which must be followed to provide a good audit.

Whatever you think about the axioms in this book, some of these axioms are connected with rules which need to be followed, and some are just statements which need to be accepted by the Regulator and by the company as an audit policy. This audit policy will help the Regulator to establish good regulation and good Oversight of companies in the "fair game" which they "play" with the companies in the area of safety.

The Regulator and the company must build respect for each other through their ongoing cooperation and this "fair game" is product of this respect. By the term "fair game," I mean: Providing an impartial, balanced, and realistic audit as part of the overall oversight process which will contribute to improvement of safety. This "fair game," as I have said before, must be a Win-Win situation: After the audit, both the Regulator and the company must feel good.

Following the axioms is obligatory: I do believe that neglecting any of the axioms will endanger the overall process of auditing.

2

Clarifications of General Terms

2.1 Introduction

There are terms and things which each person must understand to be a safety auditor. Generally speaking, the auditor is not a guy with huge knowledge or experience in particular area. It is a guy who understands the Safety Management System (SMS) and who understand companies which are the subject of safety auditing. As my old professor of electronics from university used to say: The auditor does not need to be a chicken to lay an egg, but he must have the capabilities and knowledge to determine which egg is good and which one is bad!

That is the reason that I will try to provide some introduction to clarify the terms and things which will be used by safety regulatory auditors during the audits.

2.2 Management and Engineering

There is a considerable difference between Management and Engineering.

Roughly speaking, Management is the process of planning, implementing, monitoring, controlling, and managing resources in a company for achieving company goals in a particular area. All these activities are conducted by humans. Whatever resource the company would like to plan, implement, monitor, control, or direct, the company will need humans called "employees." Even in the High-Tech companies, which have extremely high levels of complex equipment and automation, the companies need humans (employees) to analyze data and/or at least to maintain the equipment.

Engineering has two definitions. One defines Engineering as the transformation of science into practice and another defines it as a way to establish, implement, and conduct manufacturing or some other type of process.

The main point is that Management is conducted by humans and Engineering is conducted by equipment. So, in the area of auditing management systems, the emphasis should be on humans, not on equipment. Equipment should be considered only in the context of human influence on the outcome of processes where equipment is used. For example, in every Risky Industry, there are strong regulations and recommendations regarding equipment performance, established through regulatory requirements regarding availability, reliability, accuracy, precision, integrity, continuity of service, etc. All these requirements are in regard to the equipment, but also, they give information about human performance in the context of equipment functioning. Checking compliance with these requirements, the auditor checks humans: Are humans capable of setting, conducting processes, calibrating and maintaining the equipment with regard to these requirements? If not, they are lacking understanding of their processes and their equipment.

Let's elaborate this more thoroughly:

I have had a driving license for 25 years and I have driven a car in different countries. I have had only three small incidents where some small damage to my car was fixed without any big problems and I can say that I am not so bad as a driver. But having a Formula 1 car will not make me competitive in the F1 championship, simply because having a good car (equipment) is not enough to produce good results (products, services). I need support expressed through other humans which will deal with settings, maintenance, race support, training, logistics, etc. So, having the best equipment is not enough for good results. Humans using this equipment need particular knowledge, skills, experience, and attitude for good results. If all those things are structured and organized in a systematic way, it means that the companies need a management system.

Generally, I can say that Engineering (equipment) in the companies is supported by Management (humans).

2.3 Hazard, Threat, and Risk

Speaking about safety, we must start from the basics.

Everybody employed in a Risky Industry must understand and differentiate between Hazards, Threats, and Risks. In SMS regulation, there is a requirement for each subject to have a process of hazard identification and risk assessment. In addition, they must produce list of hazards identified and risks calculated and assessed.

A hazard is a situation which can produce harm or death to people and damage or destruction of assets. There is hazard in driving a car (you can have a crash!), but it does not necessarily mean that it will happen every time you drive your car.

In quantifying the hazards regarding the severity (how strong the consequences are if it happens) and frequency (how often it can happen), we are calculating the Risk. In some industries (e.g., the automotive industry!),* they also use "detection" (how easily the hazard can be detected) to quantify risk.

Threat is a synonym for hazard and it is used mostly in finance, banking, and economy. From the Risky Industries, only the petroleum industry uses threat with the same meaning as hazard.

I will use the terms hazard and risk in this book with the definitions presented at the beginning of this book (Definitions of the Terms Used).

2.4 Two Types of Safety

There is something which needs to be clarified at the beginning: Risky Industries are subject to regulation which covers OHSMS (Occupational Health and Safety Management System) and SMS (Safety Management System). These two systems are different.

OHSMS is a system which provides safety in companies where the product is manufactured and in the companies (inside!) where the services are offered. This safety system can also be found under the name HSE (Health, Safety, and Environment) and it deals with the protection of the humans, assets, and environment which can be endangered during production process or during the processes of organizing services. This is safety before the product is sold or service offered.

This area is well-regulated and there are known rules on how to achieve it. For example, there are studies which provide information on matters such as:

(a) The required volume of the premises per person in offices and manufacturing premises (6 m³ per person).

(b) The furniture which can be used in offices (no sharp edges!).

(c) How many fire extinguishers shall be available per particular volume in the premises, etc.

With these types of regulations, the auditor's job is not so complex. He just needs to directly check the compliance of the company with the requirements.

But there is another safety management system: SMS. This management system is for taking care of Functional Safety. This is safety which is applicable after the product is sold and during the provision of services. Functional Safety takes care that products and services offered to the customers are safe. And this is the problem.

---

* The ISO/TC 16949 standard is the standard which applies to automotive industry. It requires every company which produces cars and/or parts of cars to implement FMEA (Failure Mode and Effect Analysis). There, the risk is presented through RPN (Risk Priority Number) which is the product of levels of severity, frequency, and detection.

Products are sold and services are offered in different parts of the world, they are used by persons with different habits and different culture, and they are used in different environments. Depending on the area of the world, there is a need to adjust characteristics of the products or services to the reality where the products will be used or the services offered. For example, the same model of car sold in Sweden or Canada will have different characteristics from the car sold in the Middle East. In Sweden and Canada, the car must be equipped with winter tires and a good airconditioner with a good heater, and in the Middle East, the car should be equipped with summer tires and with a good air-conditioner with a good cooler.

These differences in environmental conditions, together with other differences regarding the nature of the customers, create huge problems for Regulators: They simply cannot calculate probabilities for all situations which will produce different hazards with different risks. So, the situation with Functional Safety is different. The companies in the Risky Industries are requested to identify by themselves the hazards and to calculate the risks regarding their products and services. This will help them to improve safety issues with products and/or services which can arise from their use in different environments.

Axiom 1: Whichever type of safety is under consideration, the company has the primary responsibility to provide for the safety of their workers, customers, environment, premises, operations, activities, products, and/ or services offered!

From the point of view of the auditor, Functional Safety complicates his job. He must be familiar with all aspects of Risk Management* to understand what is going on with the company's SMS. That is the reason that the overall process of auditing, especially in Functional Safety, has two phases: Documentation Audit and On-Site Audit.

2.5 Quality versus Safety

In the Preface, I mentioned that I did not understand why we need an SMS when we can implement a QMS (Quality Management System). This was a question which bothered me at the beginning when I started to work in the safety area. After a while, as I progressed with my knowledge and my

---

* Roughly speaking, Risk Management can be explained as a congregation of processes for: hazard identification, quantification of hazards to get risks, risk elimination and/or risk mitigation, preventive and/or corrective actions, sharing of safety-related information, etc.

understanding of safety, I realized why we need an SMS beside a QMS. Let's explain this with a simple example:

You have a car. On the internet, you may find information that, depending on average pressing of the brake of your car, the average time to change brake pads is between 25,000 and 65,000 km. Let's say: The brake pad on your car needs to be changed each 40,000 km. Assume that your car has driven just 10,000 km, but there is mistake in the material used to produce the brake pad. You enter your car to go for a drive, but you are not aware of that mistake (even the manufacturing company is not aware of that!), so you drive your car at a speed of 100 km/h. Suddenly, there is another car which enters highway at a bad place and is immediately in front of you. You press the brakes, but the brakes do not work because at this moment the brake pads just disintegrate. The chances of casualties in the following crash are very, very high.

Bad material for brakes is a quality issue, but the consequences in this simple example are safety consequences: Your car and the other car will be very much damaged. You and another driver will experience bad injuries (even death is possible) and any passengers in both cars are also in the same danger.

That is the difference between Quality and Safety! If the company which produced the brake pads did a Risk Assessment (what and how it may go wrong?), they probably would note this mistake. They would certainly do additional tests on the material used for brakes and they would probably change their procedures on how to test material for brakes and which type of material shall be used for brakes.

So, the difference (and connection!) between Quality and Safety is Risk Management. The quality failures may cause safety consequences and it is not strange that in plenty of books the first measure to improve safety is emphasized: Improve quality!

Anyway, there are much bigger similarities than differences between these two management systems:

a) Both of them need supporting documentation, manuals, procedures, records, management commitment, etc.

b) Both require the appointment of a responsible manager (Quality and Safety Managers).

c) Both use the same methods or methodologies (FMEA (Failure Mode and Effect Analysis), FTA (Fault Tree Analysis), ETA, Bowtie Methodology, etc.) for assessment.

It is not strange that in the new ISO 9001:2018, there is a requirement for the companies to implement "Risk-based thinking" to improve quality.

I do believe that in the future, due to similarities between Quality and Safety, there will be integration of these two management systems. Actually, in Risky Industries, there are some organizations and airlines which already have introduced an Integrated Management System, where the QMS and the SMS are integrated in one integrated manual. In the time when I was writing this book, you can find the following text on the EASA (European Aviation Safety Agency) website:

> The EASA Integrated Management system (IMS) is designed as to ensure that any additional requirements prescribed by the EU regulatory framework as well as those set forth by international agreements (e.g., ICAO standards on safety program and safety management system) in the field of aviation safety and environmental protection are taken into account.
>
> The EASA IMS is a single integrated system to manage the totality of the Agency's processes in order to meet the organization's mission and objectives. The EASA IMS will also ensure that quality of services delivered by the Agency equally satisfies the stakeholders without compromising on safety or environmental protection.

At that time, on IATA (International Air Transport Organization) website, you could read:

> With the escalating global demand for air transport, efficient operational quality and safety management are paramount. By integrating quality and safety management systems, industry stakeholders will realize considerable efficiency improvements and other benefits.
>
> With the IMX software, we offer airlines and ground service providers a simple, efficient and cost-effective quality and safety management solution.

And also at that time, on the IAEA (International Atomic Energy Agency) website you could read:

> Applying integrated management systems for nuclear facilities and activities leads to more efficient and effective nuclear energy production, participants at a recent IAEA meeting heard.
>
> A number of national and international quality assurance and management standards are being used in the nuclear industry, such as ASME NQA-1-2015 and ISO 9001:2015, to complement IAEA Safety Standards and related requirements.

The IAEA have integrated a few of the management systems. In their document "The Management System for Nuclear Installations" (IAEA Safety Guide No. GS-G-3.5), on page is written (emphasis added):

> 1.4 The objective of this publication is to provide recommendations and guidance supplementary to those provided in Ref. [2] for establishing, implementing, assessing and continually improving *a management system that integrates elements of safety, health, environment, security and economics.*

The main point with the IAEA and safety auditing in the nuclear industry is that it is done through quality auditing. On p.8 of the document "Manual on Quality Assurance Programme Auditing" (IAEA Technical Report Series No. 237) the following is written:

> The present Manual on Quality Assurance Programme Auditing contains supporting material and illustrative examples for implementing requirements for quality assurance programme audits as stated in the Code of Practice on Quality Assurance for Safety in Nuclear Power Plants, IAEA Safety Series No. 50-C-QA, Section 13, and additional requirements and recommendations presented in the Safety Guide on Quality Assurance Auditing for Nuclear Power Plants, IAEA Safety Series No. 50-SG-QA10. This Manual is directed primarily towards quality assurance programme auditors and managers and it presents methods and techniques considered appropriate for the preparation and performance of audits and the evaluation of results.

So, in the nuclear industry, the similarity between Quality and Safety is recognized on a higher level than in other industries. There, the SMS is presented through all aspects of the QMS, and the regulatory safety audit is done through an Integrated Management System.

Regarding integration of Quality and Safety in the petroleum industry, there is beautiful article in the *Journal of Advanced Review on Scientific Research** from 2015 under the title "A Review of Integrated Management System in the Offshore Oil and Gas Industry," where the present situation is given. I recommend reading it.

If the companies in these industries (aviation, nuclear, and petroleum) have integrated QMSs and SMSs, there is reason to expect that other companies in other Risky Industries will do that also in the future. And there is nothing wrong with that: If the company thinks that it will bring benefit to their operations, they are allowed to do it.

The auditor can find the QMS and SMS which are integrated and implemented as one system in some of the companies which are the subjects of an audit, so he must be familiar with both. In such cases, it is most important for the auditor to understand interactions between Quality and Safety inside the company. He must have in mind that poor quality is endangering safety and this fact should be included in the list of identified Hazards within the SMS of the company. Having in mind that the subject of his audit is SMS, he must accept the fact that some of the procedures and documentation are produced with the intention of complying with requirements for the QMS also.

---

* A. Abdul Kadir, S. Sarip, N. H. Nik Mahmood, S. Mohd Yusof, M. Z. Hassan, M. Y. Md. Dau, and S. Abdul Aziz, "A Review of Integrated Management System in the Offshore Oil and Gas Industry," *Journal of Advanced Review on Scientific Research* ISSN (online): 2289-7887, (2015) 12(1): 11–25.

2.6 Safety versus Security

From the human aspect, safety deals with human unintentional Mistakes/ Errors which can cause incidents or accidents, and Security deals with human intentions to produce incidents. Safety issues are handed by safety managers (departments) and Security issues are handled by security managers (departments) inside the company. If necessary, the police (or in some States, the army) are called for Security incidents.

Safety and security are very much different systems regarding human intentions, and they cannot be monitored and controlled in the same way and in the same department. So, although Quality and Safety Management Systems can be integrated, the Safety Management System and the Security Management System should not be integrated.

However, if the company decide to integrate these two systems, the Regulator must accept the company's decision.

2.7 Oversight versus Audit

There is a difference between Oversight and Audit.

In English dictionaries, you may find two meanings of the noun "Oversight." The first one is to making a mistake because you did not notice or you did not pay attention to some elements or complexities of the process or operation. Obviously, this meaning does not accord with the purpose of this book. But "Oversight" is defined also as the activity of gathering and providing information about the performance of a system, equipment, activity, task, etc. For the purposes of this book, this second meaning will be used.

The main point here is that "Oversight" consists of few joined or separate activities used to get a clue what is going on with the system, equipment, activity, etc. Driving a car is an oversight activity which consists of two operations: I am monitoring the road and I am controlling the movement of the car in accordance with the situation on the road (cars and pedestrians around, bending of the road, traffic lights and other traffic signs, etc.) by applying appropriate commands to the steering wheel, throttle, and brakes.

In the area of safety, oversight activities are: Constant monitoring, audits, surveys, occurrence reporting (voluntary and mandatory), investigations, surveys, and studies.

Constant monitoring,* audits, and investigations are tools which have to be independent and impartial from company influence. Surveys and

---

* It applies to traffic control and the nuclear industry where cameras provide video of what is going on in the area of interest (crossroads, stations, nuclear reactor, etc.).

occurrence reporting are based on the company view of events. Studies can be done by the company itself (a Safety Case), by an independent body engaged by the company (consultancy), or by a Regulator (special report).

By comparing the outcomes from audits, investigations, and studies with surveys, occurrence reporting, and other studies, the Regulators can get real picture of the safety performance of the companies.

As can be noticed, audits are only one method (activity) to provide information about compliance of the companies with any type of regulation. But if the audit is conducted correctly, it can provide in advance an excellent overview of what is going on in the company. It is preventive, so it can really improve the safety system before something bad happens.

2.8 Inspection versus Audit

There is big confusion in industry regarding the two terms, Inspection and Audit.

Inspection is the word which started to be used in the time when Quality Control (QC) and Quality Assessment (QA) were introduced to industry in the late 1960s. As the word is used in Quality Control, an inspection deals with measuring (inspecting) the quality of the product or service offered.

It means that the inspection is mostly manufacturing or service oriented; it is dedicated to the quality of the products or services and it is used internally, inside the companies. Inside the company, there is usually a Quality Department where inspectors inspect the products' compliance with tolerances (or services' compliance with the objectives or standards).

In general, audit is something different. It is mostly connected with Oversight of the humans or activities inside the companies by someone else.* So, the company (internally) can inspect the products or services, but the Regulators use the audits for checking the management systems and other types of compliance by the company with the regulation.

This difference is actually indicated by using different methods during inspections and during audits.

For inspections of products, the company uses metrology.† The characteristics of the product are measured by some measurement system and the results of these measurements are compared to the tolerances for particular standards for these characteristics. For inspections of services, a company uses surveys which are subjective opinions from different types of users of the company's services. These users are usually asked to fill some type of questionnaire

---

* In a few places on the internet you can find a definition regarding the audit as "independent examination" which I like very much!

† Metrology is a science dealing with any type of measurements.

to express different levels of satisfaction or dissatisfaction with the service offered. The point is that these standards used for inspection of the product or services are usually voluntarily accepted by the company and mostly they are not requested by some particular type of regulation. However, there are some types of regulation which are connected by particular standards, and companies must use these standards to be compliant with the regulation.

Methods used for audits are explained in this book. The audit methods are not connected to use of some piece of equipment or tools (instrumentation), but there are trained people (called auditors) who check the compliance of particular management systems with particular requirements in regulations.

However, sometimes, during the process of auditing, the auditor uses a process called "technical inspection," where (visually or by use of measurement systems) the performance specification of equipment is measured and compared with technical specifications related to some required standard. This process is called Verification.

In general:

a) Inspection deals with characteristics of products and services, and the audit deals with performance of processes (procedures, activities, tasks, etc.) in the companies.

b) Inspection is done by the company itself (internally), and audit is done by an independent body (a consultant or Regulator).

c) Inspection is focused on particular characteristics of product or services and audits are aimed widely at the overall performance of the company.

2.9 Compliance versus Conformance

There is much similarity between "compliance" and "conformance." To be honest, in standardization of management system's regulations, "conformance" was the word which was mostly used in the past. The ISO 9001:2015 uses "conformity" instead of "conformance," so "conformance" is an obsolete term in the standard today.

Some of the international organizations producing regulations started to use "compliance" a long time ago and it was strictly connected with regulation. Today, this term is very much in use in SMS regulation in aviation. So, in general, today "Conformity" is used for conforming with the requirements from company's management system, and "compliance" is used for complying with regulatory requirements.

The nature of the book requires the use of the term "compliance" and that is the reason that I will use it. Using "conformity" will not change anything that is mentioned in this book, so although I am recommending the use of "compliance" and "non-compliance," you may use "conformity" and "non-conformity" also.

2.10 What Is a Management System?

If you search on Google "what is a management system?," you will get around 3.7 billion hits. You cannot read all of them, but if you just read 5–6 of them, you will find different answers.

Starting from "a set of policies, processes, and procedures," continuing with "way in which an organization manages operations," and going on with "a method which provides functioning of companies," you can find plenty of different definitions. Most of them apply to the meaning of management system, or at least they cover few aspects of management system. The main point is that almost all of them are acceptable for me and should be acceptable for you also.

I understand each management system as a set of procedures which explain how employees deal with the processes in companies. And, bearing in mind that there are plenty of different processes in each company, there are plenty of different management systems: Quality Management System (QMS), Safety Management System (SMS), Security Management System (SeMS), Environmental Management System (EMS), Occupational Health and Safety Management System (OHSMS), Financial Management System (FMS), Documentation Management System (DMS), Production Management System (PMS), Logistic Management System (LMS), etc. In general, any process in any company is associated with a particular management system. Each of them can be integrated with others or they can be independent.

Axiom 2: The particular management system is built by the company, and if there is regulatory requirement to have such a system, then the management system must be approved by the Regulator through the audit or through accepting the Certificate of Compliance from any approved Certification Body!

The thing which is important to understand is that, in general, each management system consists of three constituents: Equipment, Humans, and Procedures (Figure 2.1).

As can be seen from Figure 2.1, all constituents are strongly dependent on each other, and they also interact with the external world and depend on the market situation.

If you try to establish a company, depending on the nature of the company (production or services), you will implement most of the above-mentioned management systems. The main point is that even where the two companies in the same industry, they will implement different management systems. They will differ in the structure or in other words: There are no two companies in the world which have implemented the same management systems. Let's be clear: Each company will implement a QMS, an FMS, and a PMS, but these systems will have only the same name. The management systems will

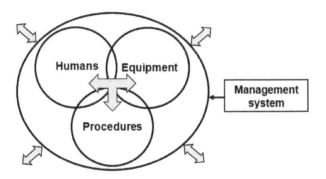

FIGURE 2.1
Constituents of each management system.

differ in equipment, humans employed, and procedures implemented. And there is nothing wrong with that!

Axiom 3: There are no two equal management system in the world! If they are the same, one of them is a COPY of the other and this will not work!

As part of each management system, there is need for a dedicated person called a manager whose sole responsibility is to implement and maintain the management system. In addition, there must be particular management system documentation (mostly known as a Manual) where a description of the overall management system, together with all system procedures can be found.

I will stick with SMS in this book, but whatever is mentioned here can be easily transformed in the areas of other management systems. You just need to take care of the specifics and requirements of each management system.

2.11 Establishing a Management System

In general, there are two types of companies in the industry: Some are producing products (manufacturing industry) and some are offering services (service industry). For example, the nuclear industry produces power (a special kind of product!), pharmaceutical companies produce drugs, the aviation and railway industries are offering transport, etc. Each of these two types of companies could be part of particular SMS regulation.

Let's explain the building of management system by one simple example from production industry. The similar example will apply also for any type of company which would provide services.

If I have designed some product and I believe that it will be successful on the market, I will try to establish a company which will produce this product.

I know what I would like to produce and I look for machines (Equipment!) which will be used during the production process. First, I will look for the premises which will be used to install the Equipment and they will be my production premises. I will look for the machines and my choice will be compromised by my wishes and the money available for my factory. I will also take into consideration the complexity of the production process which will be a function of available machines and the availability of humans which would use these machines to produce my product.

After buying the machines (Equipment), I will put my efforts into finding people who will use these machines to produce my product(s) in my company. Depending on the type of equipment bought, I will employ different types of people (Humans):

a) For complex machines which will be used in complex process with automation, I will need more educated, more skilled, and more experienced people.

b) For simple machines (used for simple processes), I will need less educated people without as much experience and with skills which could be easily learned.

I employ the people (Humans!) which I need, and next I should provide the structure and organization of my company, or, in other words: To build my management systems (system procedures) and my production system (operational procedures)!

Whatever type of management system it is, it will be built by Procedures!

And whatever the Procedures are (System or Operational), I will provide training for the Humans on how to behave in my company (System Procedures) and how to produce the products using my machines (Operational Procedures).

2.12 Understanding Procedures

As mentioned in the previous section (Section 2.11), there are roughly two types of procedures in each company: Operational Procedures and System Procedures.

Operational* Procedures are procedures which connect Equipment and Humans and they are part of the PMS (Production Management System). These procedures explain how the Humans (employees) will produce, measure, analyze, improve, and control the parts of the products which are intended to be produced. The PMS is usually a company secret because there are particular procedures dealing with processes and there are particular technologies which are used there. All of that cannot be shared with other companies. If I

---

* In some of the literature, you may find the name Process Procedures.

share my "secrets" about the technology and processes in my company, some-one else can build the same company and take my part of the market together with my profit. So, if there is any regulatory requirement for audit of the PMS, the auditor and Regulator are obliged to keep the secrets regarding the PMS.

System Procedures are the procedures which build the particular man-agement system. These are procedures which explain the structure, orga-nization, and activities to implement, maintain, monitor, measure, control, modify, and change a particular management system. For example:

a) The procedures built for the FMS explain how the company deals with their financial activities.
b) The procedures built for the QMS deal with all activities which pro-vide good quality products and services.
c) The procedures of the SMS explaining how the company provides monitoring, measurement, and control of Functional Safety regard-ing the products and services offered.

Axiom 4: When the auditor is auditing the Management System, he is checking the existence, implementation, performance, and maintenance of System Procedures!

2.13 Records

With a company that intends to be approved (certified) for a particular man-agement system, it needs to produce at least a management manual. If it is approved (certified), later the company must provide records to prove that the company is using this management system. Previously, the ISO 9001 (which is the most used standard for QMS!) used the term "records." In the new version of ISO 9001:2015, this term is changed to "documented informa-tion." Having in mind that this is a book for auditing Safety Management Systems, and the term "record" is still present there, I will use it.

Usually the records are documents which are gathered and stored by the com-pany (in writing or electronically). They are later used to prove that the company is maintaining the particular activity or procedure for the SMS. These records, to maintain their validity, need to be dated and signed* by a responsible person.

Let's mention some simple examples of what records include:

a) List of identified hazards and calculated risks of each of them (this will prove that the company is aware of the hazards and risks in its operations).

---

* In the case of electronic records, the electronic signature should be provided.

b) Alarm Logs and Logs* of Failures of operation or faults of equipment (this will prove that they register any abnormality of performance in their company).

c) Reports on all implemented preventive and corrective actions to improve safety in the company (using them, the auditor can check if the procedure for these actions has been followed).

d) Regular reports to the Top Managers on the performance of the SMS (this can prove that Top Management is following what is going on in the company in the area of safety).

Which record will be produced and stored is something which the company will decide, but there are cases where the Regulator puts into the regulation which records shall be produced and stored. The records of all safety events in the company must be stored, kept, and disseminated to the Regulator at his request or periodically, and this requirement could be included in the regulation for Risky Industries.

It is not necessary that the record must be a document. The company shall produce whatever it thinks is suitable as evidence that they are maintaining their management system. The problem is whether this shall be acceptable to the Regulator. So it is very important for auditors to be familiar with operations and technologies used by audited companies so as to have the capability to judge the validity and integrity of the records offered by the company.

In general, records are a very important category in building a management system. They should be chosen in such a way as to provide proof:

(a) To the company itself that they are following the procedures and rules written in the SMS Manual (this is something which is used for safety assurance!).

(b) To the Regulator that they are in compliance with regulation (No records—No approval!).

(c) To the public that the company is safe and it can be trusted (this is important, especially for companies which offer services, for example, airlines).

From my point of view, (a) is more important than (b) and (c)!

Company should provide procedures in their documentation on how they will manage records. The important thing with the procedure for records is that it must explain how these records are produced, who is the owner (person responsible for producing the records), who is in control of the records,

---

* Each piece of equipment in Risky Industries will usually have considerable capability of self-monitoring and fault-finding, so these logs can be found in the computer of the Control and Monitoring part of the systems/equipment controlled. So, the auditor need just to ask for printouts of these records (logs)!

and where they are stored. One of the important pieces of information is also how long the company will keep the records. In general, the company will decide how long they will keep the records, but if there is a regulation requiring a particular period (one year, three years, five years, etc.) of keeping records, the company must accept this period or they can exceed it (which is good!).

Records are not only important from the aspect of proving that the company SMS is functioning properly. A good database with records is extremely important for the company in the case of incident and accident. In such a case, where the immediate investigation is triggered, the records can save (or "kill") the company. So, whether you like it or not, records are more than just a regulation requirement: They are legal documents!

2.14 AMC

There are regulators, usually international ones, which provide additional guidance materials together with regulations which can help companies to understand and to satisfy regulations. These documents are known as AMC. The problem with the name is that AMC may be in two forms: **Acceptable** Means of Compliance or **Alternative** Means of Compliance.

In the case of Acceptable Means of Compliance, the ways of satisfying the regulation inside the AMC are obligatory. The Regulator in this type of AMC is stating that *only* these ways of satisfying the regulation are accepted by him. I strongly believe that there is some kind of reason to have these AMC to be obligatory, but I do not appreciate these Acceptable Means of Compliance.

I do not appreciate them because, although they are guidance materials associated with the regulation, the rule requiring acceptance of only these means of compliance makes them part of the regulation. It shows the lack of understanding from the Regulator side of what is the regulation and what is guidance material. I can call it "bad regulation," because it limits innovations and research efforts from the companies to find different ways (tools, methods, methodologies, etc.) to satisfy the regulation.

In such a case, the auditor does not have too much work to do: He just needs to check is that what is implemented by the company is written in the AMC.

The other AMC is Alternative Means of Compliance and this is true guidance material. These alternative means are really helping companies to have a good understanding of the regulation and to help them with the solutions. Their value is that they are showing a direction which the company may take to satisfy the regulation. From the viewpoint of the auditor, it could make them issues to realize what is solution and if the offered objective evidence regarding this solution is good or bad. Maybe it looks like a problem, but for good and experienced auditor, it is a piece of cake. Audit, as a process, is actually very simple: You ask for something and company offer the response to you! Your job is just to check if it is OK (or not OK).

2.15 Validation versus Verification

There is significant confusion in the literature and in reality on how to use Validation and Verification. Going on the internet and checking business, industry, and other dictionaries, you may notice different definitions regarding Validation and Verification. For the purpose of conducting regulatory audits on the SMS in Risky Industries, I will provide here two definitions which are proven in reality:

1. If the auditor or inspector needs to check, test, or approve some product, service, and/or management system for which specifications are requested by some international standard or by some regulatory requirements, then he is doing Verification.

2. If the auditor or inspector needs to check, test, or approve some product, service, and/or management system for which specifications are not requested by some international standard or by some regulatory requirements, but which are actually a product of the design of the company itself, then he is doing Validation.

It is not necessary that Verification and Validation are done by auditors (Regulators). Before the audit, the company can do it by itself to assure themselves that they are compliant with the regulation or the specifications which are part of their design requirements.

Following the above-mentioned definitions, it is clear that during safety audits, the auditors are doing verification of the regulation. But also, the Validations of something could be the subject of audits. Validation is especially used by auditors when the particular solution to some safety problem (preventive and/or corrective action) or Safety Case (for change) is submitted by the company to the Regulators for approval. Then the auditors will validate the proposed solution.

So, to summarize in general:

a) Verification is made when there is a need to check, test, or approve something where the international standard or regulatory requirements are existing.

b) Validation is made when there is a need to check, test, or approve something where the specifications are made by us, usually connected with our own wishes to design something.

2.16 Safety Case

A Safety Case is a document which is produced by companies and submitted to the Regulators wherein the analysis of safety Risk regarding a new equipment or operation and/or preventive/corrective action (for risk elimination or mitigation) is presented.

Regulations in Risky Industries (usually) request the submission of a Safety Case to the Regulators for installation of new equipment/system, implementation of new procedure, or modification* or change of particular equipment/ system or operation. And there is good reason for that! Whatever change in the company from the Risky Industry happens may endanger the safety of the operation. Or, in other words, whatever the reason for the modification or change (improving economy or improving safety) is, the company must conduct an appropriate safety Risk analysis again. It is done again because this new equipment or procedure may solve some of the problems, but may also create new Hazards or change the values of the Risks already calculated. As the outcome of this repeated analysis, the company will produce a document (the Safety Case) where all arguments that new equipment/systems or procedures are safe are presented. This analysis is done to satisfy, at first, the company itself, then the Regulators, and then (finally) the customers and the public.†

The content of the Safety Case must take into consideration, new Hazard identification, new risk assessment, risk elimination, risk mitigation, and applicable means (tools) for monitoring and control. All these things are the subjects of regulatory Oversight.

To analyze the Safety Case submitted to the Regulator, the auditor must be familiar with safety methods and methodologies and must be knowledgeable and experienced in the area (subject) of the Safety Case. There is a possibility here for the Regulator to engage or consult some specialists in the area of interest who are not employed with the Regulator. It is good if the Regulator has a pool of experts for such cases.

At first, the auditor must check the validity of submitted document, and after that, he must carry out a proper analysis of the data inside. The overall process should finish after the On-Site Audit is carried out to see if everything is implemented as explained in the Safety Case document.

Not every change should be presented to the Regulator in the form of Safety Case. Small changes which do not usually change the configuration of the system or slight changes of procedures are not subject to notification

---

* Whatever applies to a "change" in this book is also valid for a "modification," so I will use mostly word "change" in the book.

† Do not underestimate the importance of sharing information with the public! When bad things happen, the public is critical to all subjects included in the event. Having good communication with the public in advance (sharing information before the incidents or accidents) can help to relieve the pressure during investigations. Earning the trust of the public is very important and it can be achieved only by open and sincere communication.

through a Safety Case to the Regulator. The Regulator must be careful when analyzing Safety Cases and he must have a balanced approach in doing that. To refuse it, there must be good reasons, and to accept it, there must be no reasons not to. I will explain this in more detail later in the book.

2.17 Human Factors (HF)

Human Factors (HF), as a particular discipline in safety management, arose in the 1960s when enough gathered and analyzed data showed that humans contribute to incidents and accidents considerably more than equipment. Today, HF are extremely important in Risky Industries, and the companies there are obliged to pay particular attention to human behavior. In addition, the training for HF needs to be provided to their employees.

HF is actually a scientific approach to research of the aspects of human behavior which are contributing to incidents and accidents and how to eliminate or mitigate them. HF investigates the interaction of individuals with each other, with facilities, and with equipment. They take account of the influence of the working environment, personality, religion, and culture of the people to other people. The results of these investigations are used to create a safe workplace or safe products or services.

Regarding HF, there is need to be careful: Do not mix up HF with human mistakes/errors! Human mistake/error is actually an outcome from an incorrect human performance which is caused by HF.

There are plenty of such HF which determine our behavior. The main point of HF is seen in the national, local, and workplace environment and in human values built by social environment and community. But there are also individual HF which affect human behavior. For example: If the manager's position in the company is occupied by a person who does not have the education, knowledge, or skills for this position, then it is understandable that, during his daily duties, this person will experience anxiety and stress. What is worst, his behavior, depending on his personality, could be HF for others, and this is the classic situation when "bad managers can damage good employees."

It is important is to mention that in general, all of the HF may endanger any type of safety (OHS and/or Functional Safety).

Facilities and equipment where the production process is conducted or premises where the product is used are determined by volume of the work space, its design, and its ergonomics. In addition, the reliability of equipment and its maintainability affect human behavior. If electricity is lost during watching my favorite program at home, it will make me nervous. And if there is a lack of water due to a defective valve, I will want to fix it as soon as possible. If I do not have water for three days, then it is a problem which affects the quality of my family's lives. Tomorrow, these annoying situations, together with my anxiety and stress, will be transferred to my working place. My behavior and my performance caused by these situations will also affect my colleagues in the office. I will be a Human Factor for them.

Humans around us (our colleagues or our customers) also affect our behavior. Being in an environment with positive people actually motivates humans to improve their results. And, vice versa, being in an environment where people are not polite, are selfish, and lack understanding of a situation will produce a working atmosphere which will decrease the performance of the company. All these things affect human performance in a bad way: It is decreased! Stress and fatigue, if constant, will decrease human attention and increase the chances of bad things happening. In addition, they can permanently damage human health.

There are (roughly) eight Human Factors and these are:

1. Fatigue (Mental fatigue is especially damaging!)
2. Stress (Caused by private problems or problems in the workplace)
3. Alcohol and drugs (Even some medications prescribed by doctor can produce problems!)
4. Team work (Depends on the employees' culture, education, and social status)
5. Decision-making (Especially affects managers where a decision needs to be made, but data is missing!)
6. Situational Awareness (Especially for employees in the Risky Industries!)
7. Communication in the company (Caused by the overall atmosphere in the company and by employees' position, culture, education, and social status.)
8. Leadership (Bad managers can destroy good employees!)

From the point of view of the auditor, all these factors affect human performance and there is a need for the auditor to be familiar with them. Companies in Risky Industries need to pay attention to the humans, their behavior, their actions, working environment, and the resources available to execute the operations. There is a regulatory requirement in a few Risky Industries to companies to establish Human Factors program which will take care of Human Factors in the company. This program should be subject of audit by the Regulator.

2.18 Mistake versus Error

Mistakes/Errors endanger normal operations, and they are not welcomed in industry and also not welcomed in human lives! But there is a difference between these two terms. The difference is not strong, but it exists. For the purposes of regulatory auditing, I will explain the differences in this section.

Mistakes are defined as outcomes from wrong human choices. I have decided to do something and it happened that it was the wrong choice: Simply put, I did not get the outcome from my choice which I expected! So, mistakes are bad results from intentionally made human choice and the reason for making them is usually connected to a lack of proper information, a lack of understanding, or simply by wrong expectation. From the point of any industry, managers are prone to mistakes when they must make managerial decisions, often due to the lack of enough information.

Errors are connected with something which we know as a code, rule, procedure, etc. These are usually non-intentional human failures or unexpected equipment faults which produce bad outcomes. For example:

a) Although we have done a particular activity every day in the past, we can still make an error executing this activity today.

b) In regard to equipment, an error can show up in computer code when there is situation which is not covered by this operation.

c) In operational procedures, an error can show up when dealing with a situation which is not covered by the procedure.

If made by equipment, the error will affect overall operation. If made by humans, the operation can be rectified, but this is the subject of early detection. Errors made by humans are usually connected by Human Factors (non-intentional outcome due to any of HF).

Error has a more technical meaning and it is used in industry. The term "mistake" is used mostly in management. From the point of auditing, there is connection between mistakes and errors: Usually the mistakes (wrong decisions!) can produce systematic failures and auditors must understand that. Errors are mostly random.

2.19 Just Culture

Functional Safety needs data to be proactive and predictive. But the data is missing that would make it proactive and predictive.

Data is missing because people do not report safety significant events, so if we try to statistically analyze them, the conclusions are not good. Or, as statisticians would say: There is no sufficiently large sample for analysis!

There are a few reasons for not reporting the safety events. Sometimes, people think that it is not their job, sometimes they like to protect others from punishment, sometimes they are included in the event, so they like to protect themselves.

Anyway, with intention of gathering more data, the Just Culture initiative was launched in the 2000s: If you report a safety event, you will not be prosecuted or punished for what happened. But it is not quite so simple.

The breach of safety can be intentional or non-intentional. Intentional breach of safety means that humans intentionally, knowing the effects and consequences, do something which can endanger humans and assets. This is called Criminal Intent, and sometimes even it can be an act of terrorism and is processed by the police. It is actually not a safety issue, but a security issue.

Non-intentional breach of safety (known as mistakes/errors) are caused by Human Factors (HF). Human Factors are products of bad behavioral habits or situations, which, if present, can produce human mistakes/errors. Sometimes, these HF can contribute to intentional mistakes/errors (e.g., I am annoyed by some decision of the managers and I break something), but in general, they are considered reasons for non-intentional mistakes/errors.

Just Culture is a safety concept which encourages reporting all mistakes/ errors which occurred due to HF and guarantee to the employees that nobody will be punished for reporting it. For example, I slept for only three hours last night, because my child was ill. I showed up to my workplace next morning, but I could not focus and I made a mistake in adjusting the settings of a machine which produced high pressure on the tank of a dangerous chemical substance. Luckily, I noticed it and I reacted immediately by decreasing the pressure, so nothing happened. Nobody noticed it, but I will report it.

I should not be punished for my error, but it will trigger few actions.

Knowing this information about the event, the company's Safety Manager will analyze the event, and as a preventive action, he will insist on the installation of a pressure gauge with an alarm on the tank which will go on any time that the pressure is too high. It does not mean that in the future, I will be allowed to work on the process with only three hours sleep the night before. But, bearing in mind that humans are not always aware of their decreasing performance, this alarm can tell them that errors happen even when they are doing and expecting all the best.

In addition, next time, my colleagues (who were informed about the event!) and I can ask for free day(s) if we experience a similar situation. Our manager will understand what was going on, so next time, he will approve these day(s), and he will prevent HF (poor sleeping caused by a child's illness) from causing such an event.

Of course, Just Culture must be used, but it may not be abused!

The auditor must have a good understanding of HF and he must differentiate between activities in the company based on HF or a bad safety culture. Anyway, too many such events indicate a bad atmosphere in the company or bad leadership.

There is another aspect of Just Culture:

That what the companies are for their employees, the Regulators are for companies! OK, maybe it is not the same situation, but it is very similar. So, Just Culture does not apply only to the partnerships between the

TABLE 2.1

Partnerships in Regulation

Level	Superior	Subordinate
1	Company	Employees
2	State Regulator	Companies
3	International Organization	States (Regulators)

companies and the employees, but also applies to the partnerships between the States (Regulators) and the companies and to partnerships between the International Organizations (IAEA, ICAO, etc.) and the States (members).

So, there is a three-level partnership in regulations for providing safety in Risky Industries. The pairs are given in Table 2.1.

At the end of the year, the safety regulation requests all companies to provide information for safety events to the Regulators, and the same thing happen with Regulators, i.e., they must provide processed information from the companies in their States to the international organizations also. So, Just Culture must apply to all three levels.

This is very important for Regulators: They must be careful with punishment of the companies for the wrong performance. The regulation must be followed and on all three levels Superiors and Subordinates must struggle to comply with the regulatory requirements. But the most important thing is to understand the reasons behind the non-compliance. In medicine, the good doctor will try to find reason why you are ill and then he can start with the cure. If he is only dealing with the consequences and not with the causes of the illness, he is not a good doctor.

Be careful: The auditor must "read between the lines" when the HF are considered!

So, in all three levels from Table 2.1, Superiors should not just be dedicated to fines; they must "diagnose" the reasons for the "illness," and then they can consider the reasons for punishment. Auditors are on the "frontlines," so they must have a significant role in these activities, and they must especially appreciate and value the effects of Just Culture.

3

Regulation

3.1 Introduction

The word "regulation" is used mostly to explain the congregation of laws, rules, standards, or procedures which are accepted by a particular authority in the State and/or on the international level. Regulation is a legal function in the State, and entities must abide by the regulation. If they do not abide by it, they can be subject to legal prosecution.

Regulation must be controlled, and the performance of the companies under the regulation must be monitored. The body who is responsible for control of the regulation and for monitoring of abiding by the regulation is called the Regulator. The monitoring process of the companies by the Regulators regarding fulfilling the regulation is known as Oversight. To be successful with Oversight, the Regulator must have employees with a high level of knowledge and *understanding* of regulation. The main word (italicized!) in the previous sentence is "understanding," because there are a few important aspects of regulation to be understood by anyone. The Regulator especially must take care of each of these aspects. These aspects are explained in this chapter.

3.2 How to Pass a Regulation?

A particular regulation is usually made on the proposal of the Regulator.* It needs to be a process where there are particular inputs which have triggered this process (reasons why the regulation is needed!) and there are particular outputs (targets which need to be achieved and rules which need to be abided by!). The process used to pass regulations connects the inputs (reasons!) with the outputs (targets!). It is a complex process and there is a

---

* To be honest (and truly correct!): Different countries apply different way of making a regulation. But most of the countries have established regulatory bodies (Regulators) which are in charge of regulation in particular areas of life and/or industry.

need for a considerable amount of knowledge and skills to be used to produce good regulation. Regulation in the State (or industry) is describing the Regulator and fulfillment of State regulation is describing the company.

Risky Industries are usually subject to particular international regulation. The reason for this international cooperation is because the accidents and incidents can endanger areas wider than territory of one State, or the accidents and incidents can happen between the subjects of different States. For example:

(a) Aircraft (or trains, ships, etc.) from one State can crash in another State.

(b) Chemical pollution of one river (or lake, sea, ocean, etc.) can affect plenty of States.

(c) Accident in a nuclear reactor in one State can affect by radiation another State(s), etc.

In the scope of this international regulation, the implementation and maintenance of the regulation must be done through implementation and maintenance of a particular Safety Management System (SMS), and it needs to be controlled (overseen!) by the State Regulator through audits* at regular intervals (every six months, every year, or every two to three years). An aim of the regulatory audit of any[†] management system in each company is to gather data which will prove that the company complied with the (any) regulation.

Unfortunately, the regulation is not always clear and precise in its requirements, and it gives a lot of chances to the companies and to the Regulators to use or abuse the regulation. That happens pretty often when there is an International Body which is producing the regulation in some Risky Industry and the States need to transform this regulation's requirements into their legislation. The best way to do that is to produce good translation of the International Body's regulation which will be part of State legislation. But plenty of countries look on that as undermining their independence, so they try to deal with "soft" and "non-accurate" translations, often putting things in the regulation which are very confusing for the companies.

During the process of bringing a new regulation or changing the old one, there is need to have an "open channel" for communication between the Regulators and the companies in the State. The good Regulator will always ask for opinions (comments, advice, suggestions, etc.) from the companies before the regulation is passed! It is wise for each Regulator to take into consideration all these opinions (comments, advice, suggestions, etc.) from the companies, because it will help to pass good and sustainable regulation.

---

* More details about types of audits can be found in the next chapter (Chapter 4).
† This applies not only for SMS, but also for any management system (QMS, EMS, SeMS, HSEMS, etc.).

Bringing bad regulation is not only a problem for companies, but it is also a problem for Regulators!

There is a need for a balanced approach to establish regulation, particularly in Risky Industries. This balanced approach is guided by specifics of this type of industries. Regulation in Risky Industries must be built to eliminate or mitigate the causes for accidents and incidents (if there is no cause – there is no accident or incident). But having in mind that this is not always possible, the regulation must be produced also with the intention to eliminate or mitigate the consequences. It means that each regulation in any Risky Industry must provide the legal basis for using rules (tools, methods, activities, measures, procedures, controls, etc.) which will try to eliminate and/or mitigate the causes for accidents and incidents and their consequences.

Before the regulation is in force, a good Regulator should provide a workshop when all aspects and ideas about regulation will be presented and considered. Before the workshop, a good Regulator should provide the draft text of the regulation to companies, to give a chance to the companies to prepare themselves for the workshop. The workshop must be organized in two parts. In the first part, the Regulator will explain the reasons why the regulation is needed, what is required by the regulation, and what are expected effects of the regulation. It is good if the Regulator submits some analysis or data which will support the text of the regulation. In the second part, companies may ask questions and express their views of the regulation.

In the case of bringing in a new regulation or changing an existing one, the Regulator must provide enough time for companies to understand the regulation and to calculate what resources they need to fulfill the regulation. In addition, they will need time to implement the regulation, and the Regulator must be reasonable about the time needed for implementation: Not all companies can respond in a timely manner to a new or changed regulation.

Another aspect of good regulation is that it must not kill innovation and new ideas inside the companies and industry. The balanced approach is also applicable there: The regulation must be strong enough to provide control over Risky Industries, but it must be weak enough to support new technologies (methods, measures, activities, etc.) and new ideas in the areas of interest.*

There is one very important side of regulation which is not always understood by the Regulator: The regulation is always a burden for the companies!

To satisfy the regulation, companies must employ additional people or must use other resources. It is always costly and it is time-consuming for them. This was noticed by some Regulators, but mostly it is noticed by politicians. They always try to promise that the volume of regulation will be decreased, and maybe it can be really done in some of the industries. But speaking about Risky Industries (as said before), the Regulator must have a

---

* This is something which applies also for Standardization! Whatever standard is produced, it must not kill development in the industry based on innovation and new ideas.

balanced approach. These industries may deal with decreasing the volume of regulation, but they must provide enough legal means not to allow the companies to endanger Humans, assets, and the environment. In addition, the Regulators must not call off from their duty to oversee the company. Simply, the Regulators must assure themselves and the public that they have control of each company in the Risky Industry.

There is another aspect of the regulation which could be connected by the oversight activities of the Regulator. The regulation can be also used to protect domestic industry. Although world standardization is very much accepted by the countries all around the world, the regulation can be used as protectionism. Bringing in a regulation which is more stringent than international one in the Risky Industry could favor domestic companies. Having in mind that these companies already comply with the State regulation, building a nuclear power plant or chemical factory in this particular State could be very costly for the foreign company (which abides by international regulation).

In a case where the company is established by foreign investors, the Regulator must be careful in auditing this company: They may apply some solutions which are good for the regulation of their State, but do not comply with the domestic regulation.

3.3 Accreditation, Certification, Licensing, Attestation, Approval

There are few processes in industry which are connected with official recognition of the companies or individuals. All these processes are conducted by Regulators or by different bodies. These processes can be classified by their purpose, so there can be Accreditations, Certifications, Licensing, Attestations, and Approvals.*

There is big confusion in reality about Accreditation and Certification. People do not always understand what is what. If you go on the internet and try to find a clear definition, you will be disappointed. Anyway, in this chapter, I will present one of the views which I do believe complies with requirements in the Risky Industries. Roughly, the Accreditation Body may accredit a company to provide Certification services (to certify other companies).

The ISO (International Standard Organization) has its own definitions of Accreditation and Certification. According to them:

---

* The definitions of Accreditation, Certification, Attestation, and Approval given in this paragraph are for the purpose of audits only. There are different aspects and different meaning of all these words, but I focused here only on the aspects and meanings which fit the purpose of the book.

- The Accreditation is a "third-party attestation related to a conformity assessment body conveying formal demonstration of its competence to carry out specific conformity assessment tasks" (ISO/IEC 17011 Conformity Assessment-General Requirements for Accreditation Bodies Accrediting Conformity Assessment Bodies).
- The Certification is a "third-party attestation related to products, processes, systems or persons" (ISO/IEC 17000 Conformity Assessment-Vocabulary and General Principles).

Accreditation is the process of formal recognition of company or individual (by a so-called Accreditation Body) that the company or the individual are capable of doing a particular job (activity, process, etc.) in accordance with standardized specifications (standards!). For example:

(a) Accredited laboratory – It is recognized that the equipment, processes, and procedures which are used there are in accordance with the particular laboratory standards in that area.
(b) Ambassador submits Accreditation Letters to the Ministry of Foreign Affairs (or to the President) of the State where he is appointed as proof that he can do his job in accordance with the Vienna Convention on Diplomatic Relations of 1961.
(c) Hospital (clinic!) is accredited for particular medical process, type of surgery or method of cure, etc.

Accreditation is usually connected with a conformity assessment. A conformity assessment is a process of using testing and inspections of products (services). These testing and inspections provide data which are used as proof that the company (or the individual) have the capability to produce these products (or offer these services) in accordance with specifications from regulatory requirements or from some international standards.

Certification shall be done by Certification Bodies (which are already accredited for such an activity by Accreditation Bodies), and Certification is the process of providing a certificate for company or individual that they have achieved something. This "something" is different for companies and individuals. For example: The companies are usually Certified that they have implemented a particular standard or they have achieved particular level of excellence. The individuals are Certified regarding reaching some qualification or some status.

There is usually a formal procedure of Conformity assessment and Verification regarding some requirements and/or standards which later produces data used for Certification. After this formal process, if the assessment and Verification show compliance, the company or the individual will get a written document with the name: Certificate. This document is an official document and it has legal validity: The company or the individual with Certificates are subject to some legal privileges and legal responsibilities in the area mentioned in the Certificates. For example:

(a) When the company reaches some milestone regarding environmental protection, the particular Certificate is issued as a proof.

(b) When an individual finishes some training, the particular Certificate of Attendance is issued for such purposes.

(c) When the company implements some management system in regard to a particular standard (ISO 9001, ISO 14001, etc.), the Certificate of Compliance is issued, etc.

There is something else which has a similar meaning to Certification, and this is Licensing. The License is also given to companies showing that they have implemented particular standard or they have achieved particular level of excellence, and to individuals regarding reaching some qualification (but not some status).

The difference between the License and Certificate in Risky Industries is that Licenses are given by the Regulators and Certificates are given by a training organization or by the company itself. In Risky Industries, Licensing and Certification is mostly used for Humans (employees in the companies). There are regulatory requirements for particular employees who are in particular positions to be Licensed or Certified.

Certification is more popular than Licensing in industry, and honestly speaking, I do not know why. My humble opinion is that popularity is higher for Certificates than for Licenses due to driving licenses. Most adults have driving licenses today, which implies that importance of licenses is not so big if everyone can obtain it. But in Risky Industries, Licenses are stronger than Certificates.

The validity of Certificates and Licenses can be limited or unlimited. For Certificates and Licenses with limited validity, there is a particular period of time for which the Certificate or License is valid. During this period of time, the company or individual can enjoy the privileges and responsibilities of the License and/or Certificate. After this period of time, the company or the individual must renew the License or Certificate. Renewal is usually done in the same process as earning the License and Certificate. It is not strange that Certificates and Licenses are renewed automatically if the person has continually worked in the position which is subject of Certificate and License. If there is a time gap when the person was absent from his duties, then the process of Certification or Licensing is repeated from the beginning.

Attestation in industry is a formal process of recognition (proof, evidence!) that the product(s) of the company satisfies the specification required by particular standard in that area. In addition, Attestation can be used also as a proof that the product is genuine (belongs to particular production process of well-recognized company). Attestation in industry is connected with products only.* There is no service which can be Attested.

Approval in industry is a formal and official process (conducted by a Regulator!) for giving permission to particular company to do its job. The State Regulator, by issuing Approval, confirms that the company were overseen in a particular area and, in their opinion, the company complied with regulations. In simple words: It is a statement of the belief of the Regulator that the company is good in some area and their performance in this area is acceptable. Approval is connected with services, activities, and processes. It does not apply to products.

Approval is similar to Accreditation, but Accreditation has a stronger legal importance. The reason for that is: The rules for Accreditation are standardized and the rules for Approval are not (each State Regulator may make their own rules).

For example:

(a) The particular design of aircraft is approved and the company may start production.

(b) The particular nuclear plant is approved for production of electrical power.

(c) The particular corrective action by the company is approved and the company may execute it, etc.

Why do auditors need to be familiar with Accreditations, Certifications, Attestation, and Approvals?

Because the auditors, doing their job (auditing) will encounter them. All the documents presented by the company to the auditors can be products of particular processes known as: Accreditations, Certifications, Attestation, and Approvals. For example:

(a) The laboratory which is providing calibration services for any company in Risky Industries must be accredited.

(b) The Regulator may accept Certification regarding a Quality Management System (QMS), an Environmental Management System (EMS), or any other management system issued by any recognized Certification Body.

(c) There is need for some of the instrumentation or sensors used in Risky Industries to be Attested.

---

* In the wider (linguistic!) meaning of the word "attestation," it is connected with a statement that something (document, certificate, product, etc.) is true (genuine) or that something is existing. For example: The Ministry of Education of one country attests the Educational Diploma (or its copy!) of their citizens as true (genuine) and the citizens can use the attested document for employment in other country.

(d) Approval of a company is a consequence of the compliance checked during Regulatory Audit. In addition, if previously Approved, the company may use this Approval to ask for renewal.

So, it is important for the auditor to be familiar with all of them. Let's summarize all of them in a simple way which will fit the audit's purpose:

(a) The Accreditation is proof that company can do a particular job (operation, activity, task, process, etc.).
(b) The Licensing and Certification is proof that a company or employee has achieved a level of something (expertise, status, knowledge, education, etc.).
(c) The Attestation is proof that a product (document) is true or genuine.
(d) The Approval is proof that a company was assessed by a Regulator and this assessment produced the belief that they can do the particular job.

3.4 Standards versus Recommendations

Each regulation consists of many clauses and paragraphs, but not all of them have the same significance. This is applicable mostly for international regulation, because international organizations are also trying to give a chance to new companies. Starting a company in a Risky Industry is very expensive, so usually there are two types of clauses and paragraphs inside regulation. It could happen in each industry and could cover a particular period of time, especially when the new technology is introduced and it is not covered by present regulation.

I know that this is common in aviation: All documents which are part of aviation regulation are named Standards and Recommended Practices (SARPs). In the regulation, the standards are represented by paragraphs and recommended practices are represented by recommendations. Paragraphs for standards are something which *shall* be fulfilled. Recommendation for recommended practices is something which *should* be fulfilled. Actually, the point is that in fulfilling the standards, you are providing safety, and in fulfilling recommendations, you are putting in a bigger margin of safety.

For the purpose of auditing compliance with the regulation: The paragraphs (standards) are stronger, because everybody shall fulfill them, and recommendations are weaker requirements and they, if not fulfilled, are not a matter of non-compliance.

3.5 Regulation and Regulators

Regulators take care of how the regulation is to be shaped and to be passed, and later, they oversee the implementation of the regulation by the companies. But, if something bad happens, they are not included in the investigation process as investigators. The reason for that is because they are a vital part of the regulation and as such, they will be investigated.

It means that:

(a) If the regulation is not good (It allows accidents or incidents to happen!), this is the responsibility of the Regulator.

(b) If the regulation is not followed by the companies, there is again the particular responsibility of the Regulator. (The Regulator must provide implementation of the regulation by the companies!)

(c) If the Regulator does not oversee the companies, it is a problem! (The independent and impartial control over the safety performance of the companies is missing!)

So, Regulators are in charge of regulation, but if an incident or accident happens, they are part of the problem, which means they are also subject to investigation.

There are different types of bodies which are responsible for the investigation of incidents and accidents in Risky Industries, and usually they are independent bodies which have nothing in common with the Regulators.

The typical examples for this are the FAA and NTSB in USA. Federal Aviation Administration (FAA) is the US Regulator in the aviation area and the National Transportation Safety Board (NTSB) is the independent investigation body. The FAA is responsible for national regulation in the area of aviation and they provide Oversight of all aviation subjects, but if an accident happens, the NTSB is in charge for the investigation and they will also investigate the level of involvement of the FAA regulation (or FAA actions!) in causing the incident or the accident.

Maybe this reality is another call for auditors to put more efforts to provide good audits, because a poorly conducted audit may not find hidden deficiencies and people can die and assets can be destroyed. In such cases, the auditors (the Regulator!) will not avoid the investigation.

But there is something else which each Regulator must have in mind. There is an adage in Western civilization which says "It is not good in life to be a bigger Catholic than the Pope!"

What does it mean in the scope of regulation?

It means that the companies in the Risky Industries cannot reach a higher level of safety performance than one which is promoted by their own Regulator! So, if the Regulator is good (knowledgeable, competent, open-minded, etc.), he can also look for these attributes from the companies. If the Regulator is neglecting his duties and his responsibilities and if he is not professional, the companies will do the same. In general: If the Regulator is not looking for compliance, the companies will not provide it.

The Regulator's employees (by the power of law!) have power over the destiny of the companies, but this power must be used for good and not abused. The Regulator must be a "driving force" in achieving safety, and the employees inside must be a "specimen" which will be used by the companies as a model of how their employees need to behave.

3.6 "Grey Area" of Regulation

There are many "grey areas" of regulation in Risky Industries which are good for the Regulator to consider, but I will focus in this book only on two of them. By the term "grey area," I mean the areas which cannot be regulated well, but which have a considerable effect on safety. These are understaffing and finances. Understaffing will be considered later, in Section 10.7 (There Is Not Enough Staff in Company to Maintain Safe Operations), but this is a good place to discuss financing.

It is clear to anybody that lack of money can affect humans. In today's world, this is probably most often the reason for stress in human life. Having in mind that humans are in the companies, if the company is in bad financial situation, it will for sure affect their operations.

There is a beautiful article named "Financing Constraints and Workplace Safety"* from J.B. Cohn and M.I. Wardlaw. Although this is an article dealing with Occupational Health and Safety, it totally applies to Functional Safety also. In the article, the authors present data from their research regarding influence of the bad and good financial situation on incidents (injury rates) in the company. The results from the research shows a high correlation of the employee's injuries with the status of company's financial situation. The data shows that the number of injuries in daily operations increases with a bad financial situation and decreases with a good financial situation.

This article proves something which was also logically considered in the past. Obviously that bad financial situation is one of the hazards in the companies in all industries. Our concerns are Risky Industries, and the Safety Regulator must be aware of that.

Nevertheless, in some industries (aviation!), some regulators have a requirement for economic audits of the new companies (airlines!) which is

---

* Article first published online on 14 September 2016 in *Journal of Finance* DOI: 10.1111/jofi.12430 (https://faculty.mccombs.utexas.edu/jonathan.cohn/papers/injuries_2016mar_public.pdf).

expressed by the amount of money which they need to invest in establishing the company; this is a "grey area" for most of the Risky Industries.

Imagine that the Regulator oversees the company in a Risky Industry. Through some of the oversight tools (obligatory reporting system, complaints, audit, etc.), the Regulator gets information about increased safety events in the company. From other sources (stock exchange, financial report, media, etc.) the Regulator understands that this company has financial problems. The correlation is here, the risk of incident and accident is "in the air," but the only thing which the Regulator can do is to increase the oversight activities over this company. The Regulator, in such a situation, may not close the nuclear power plant or stop the airline from operating, even though the possibility of an accident is high.

The reason for that is that the correlation of the financial situation and the number of injuries is only a statistical operation and it cannot be used as objective evidence. "Statistical" means that, on average, there is an increase in injuries with a decrease of the cash flow in companies. But not all companies experience such a strong correlation. There are some of them which are not affected by bad financial situation, but having in mind the calculation is included into the average, obviously their number is not so big. These companies use savings measures in other areas, but the safety area is untouched. And having in mind that the Regulator cannot know how the company will perform in the future, the only thing which can be done by the Regulator in each "grey area" is to increase the frequency of oversight activities in such a company.

Increasing the oversight activities in the company which is in financial trouble should be done in a balanced way. The company is already in trouble and the increasing Oversight can add additional pressure on it. So it is wise if the Regulator organizes a meeting with the management of the company and explain what they must do, but in general, the Regulator will support company activities to solve their problems.

That is what partners do!

3.7 Regulation and Auditors

There are good reasons that the auditor must be familiar with each regulation and must have a clear understanding of the requirements. Most important is that he must have an understanding why the particular paragraph is part of the regulation and what are the consequences if this paragraph is not satisfied. Having this understanding does not mean that auditor will not use the regulation during audits! It does mean that he will have a clear picture of what is requested from the company and he can successfully check that the company satisfies it. Do not forget: The reason to implement SMS is to produce a system which will eliminate or mitigate causes for accidents and incidents and particular safety consequences affecting Human lives, assets, and the environment.

From the point of connection between audits and regulation, the auditor must understand that there are many ways to satisfy regulation. If there is only one way to satisfy it, then this way must be clearly stated in the regulation. As help in satisfying the regulation requirements, a lot of States or organizations provide guidance material in the form of Acceptable (Alternative) Means of Compliance (AMC),* but even in these cases, the auditor must understand that company is the subject which decides how to satisfy regulatory requirements. Satisfying regulations means that company complies with the requirements from the regulation.

Axiom 5: There are always many ways regulations can be satisfied by the company. The auditor must understand how the company is trying to comply with the regulatory requirements to be able to audit the company!

The understanding of the way how the company has satisfied the regulation is essential to the successful conduct of an audit. Simply, auditor may not (cannot!) audit something which he does not understand!

3.8 Post-holder

The CEO, General Manager, and/or Safety Manager in the companies of Risky Industries are very important. These are not only positions which need managerial expertise, but also, they are based on particular knowledge about the organization, technology, and safety of the processes (operations, activities, etc.) inside that company.

Post-holder is a position in the company which can be given to the person who is the subject of approval from the Regulator in this particular industry. The company submits the CV and other documents (education, knowledge of foreign languages, court certificate, visas, etc.) of the person they would like to put in this position to the Regulator. The Regulator will go through the submitted documentation, they will have an interview with the candidate, and they will approve or disapprove the person. If the person is disapproved, he or she cannot hold this position. In such a case, the company must nominate another person who is also subject to the Regulator's approval.

The Regulator has the power to decide if the particular person will govern or manage the company, but it should be done in an appropriate way. This appropriate way is to put qualities which this person must have into the

---

* This is the subject of Section 2.14 (AMC) in the previous chapter (Chapter 2).

regulation, and it must be done in a balanced way. The meaning of a "balanced way" is to put requirements regarding the Post-holder position in the regulation, but it must be done in general: Do not provide too many details about the requested qualities. Too many details mean:

(a) Do not put a particular area of education, mention just "high education," "master's," or "PhD."

(b) Do not put a particular high level of experience. For example: Putting a requirement for ten years' general experience and five years in this Risky Industry.

(c) Do not put extra requirements, instead request only "the company (candidate) can submit whatever they think will improve the chances for this candidature."

(d) Do not put too many languages as a request; instead, ask for working knowledge of English language. The Risky Industries are usually international, so everybody there should be excellent in English.

(e) Of course, put the requirement that the candidate will be subject to interview with the Regulator! Whatever documentation is submitted, the interview can provide chance to evaluate the person during his "live performance."

The reason for such a general regulation about Post-holders is that you cannot measure a particular attitude to safety or expertise in the area based only on documentation. This is something which is embedded into the personality of the person, and documentation cannot provide right information regarding that. I had a chance to work with a guy on a Safety Project in India and he was my Team Leader. He had an impeccable CV (Pilot, ATCO, instructor, etc.) and worked in many companies in respectable countries, but his safety understanding was extremely poor (unfortunately).

In general, a good manager cannot be educated! He is born! If you disagree with this, please remind yourself of Steve Jobs, Bill Gates, Mark Zuckerberg, etc. Not one of these guys have finished university, but they "put their own stamp" on business history.

Providing the general requirements in the regulation regarding the Post-holder, the Regulator provides the chance to choose a good person who can be investigated intellectually, emotionally, and whatever way is necessary during the approval process (interview, personality tests, etc.).

Choosing an inappropriate person is a liability for the company, but also for the Regulator!

As I said before, the Regulator must have a balanced approach to the approval of the Postholder. Do not forget that the company had its own reasons to propose this guy, so not approving him could cause problems. The company even may sue the Regulator. So, whatever is the reason not to approve the candidate, it must be elaborated thoroughly by the Regulator.

Eventually, the already approved Post-holder may be removed by the Regulator if there is enough evidence that he does not deserve this position.

3.9 "Personality" of Regulators

There are different types of Regulators and I would like to express my opinion about them here.

The first type of Regulator is one who does not have a proper understanding of the area which is covered by regulations. To elaborate on this, I would like to present here a simple example which is connected to the regulatory document of a Risky Industry's Regulator from a State (country) which under any circumstances cannot (and may not!) be treated as a country from the non-developed world. I found this example as instructive, because wrong understanding of regulation usually happens to Regulators of non-developed countries which are struggling to reach some level of legalization in their State Administrations.

In the official document (which is published as guidance material!) of this Regulator, in the explanation of how to employ safety personnel, it is written "employ persons with competency (*some knowledge* of the science of safety)." As you can notice in the italicized words, this Regulator (from a well-developed country of the Western world!) is recommending to subjects in one Risky Industry to employ someone with "*some knowledge*." But there is more: Going further in this document, you can read that "Appointing the good safety manager is critical." My question: How the person can be a "good Safety Manager" (a critical position!) in combination with "some knowledge of the science of safety?"

In the following pages of the document, this Regulator writes that "small companies may employ a part-time safety manager"? I am wondering: Is safety really a part-time job? I strongly disagree!

Later in the text, they explain the position of Safety Manager in detail: "The safety manager should have a sound understanding of safety management *principles*." My question: Could the Safety Manager with "a sound understanding of safety management *principles*" be capable of identifying the Hazards and calculating the Risks? Can he provide proper execution of the preventive and corrective actions?

Going further, there is an example in the document* regarding the company which has sent their Safety Officer on two courses, one course regarding Safety Incident Investigation and another course regarding Human Factors

---

* This is guidance material, so there are a few examples in the document which needs to increase the understanding of companies regarding SMS implementation.

and Error Management. And the company CEO thinks that the €4,000 given for the training are a good investment in the future!

So, is that all? Is the implementation of Safety in Risky Industries about preventing incidents and accidents? Where is the training for Hazard Identification and Risk Assessment? There is requirement in each SMS for both such things, but the "Safety Officer" is not trained for this? I have attended 12 different safety courses and I spent thousands of hours reading articles and books on the internet regarding different methods and methodologies on how to identify hazards and how to calculate the risks. EUROCONTROL in its Institute of Air Navigation Services provides 22 courses for Safety Management. But this Regulator is happy if the "Safety Officer" has attended two safety courses!

I thought that safety is primary in each Risky Industry and I wonder how it can be maintained and provided by "some knowledge," "a sound understanding of safety principles," and by "two trainings"?

I would like to add here also that I am not satisfied with the text regarding Risk Management in the document, where the Risk Management (in my humble opinion, it is the Core of Safety!) is not emphasized enough.

But, let's be honest: This document is not a bad document at all! Actually, if these basic misinterpretations were not there, I would like this document. Simply these mistakes are showing a basic misunderstanding of safety, and if the Regulator is missing such an understanding, how he can improve the understanding of the companies in the Risky Industries?

This is sad story, but if there is "science of safety," respect the word *science*! Implementing Risk Management (Risk Assessment and Risk Mitigation) needs knowledge of method or methodology, and all serious methodologies implement statistics, probability, Boolean algebra, etc. All these areas are part of the science, and they are not "some knowledge!" The average book in statistics is around 400 pages and average book in probability is around 500 pages. A Good Safety Manager must be familiar with these things or at least, the company must employ a mathematician (a specialist!) who will be trained in Risk Assessment methodologies.

In the example above, I have presented only one type of "personality" of Regulator where the misunderstanding of safety is evident. But there is more about Regulators.

The second type of Regulator is the one who has too bureaucratic an approach. I define these Regulators as ones who are "happy" with the existence of the document (procedure, record, etc.), but they do not care what is inside the document (procedure, record, etc.). These are Regulators who during oversight activities are just formally checking the documentation, and they do not go into deep analysis of whether it is sustainable and what is going on with the activities of the company. If something bad happens, these Regulators will issue the statement that "they have checked the company few times in the past period and the company complied with regulations."

A good example of such a Regulator was Mineral Management Service (MMS).* MMS was an agency of the Department of the Interior in the USA that managed the nation's natural gas, oil, and other mineral resources, and they were the Regulator responsible for Oversight of the activities on the *Deepwater Horizon* rig in the Gulf of Mexico. Although they did a lot of audits on British Petroleum rigs, they did not notice the missing procedures and quality culture in these rigs. It was systematic error; it was not individual or random error! Archived data said that BP had a record of plenty of missing things regarding the safety of their operations in the past. They even had very serious accidents, e.g., when the explosion in Texas City Refinery in March 2005 happened (15 people died and 180 were injured!). The biggest mistake made by MMS was that their auditors, instead of focusing on the *quality* of audits, were focused on the *quantity* of audits and the formal existence of the procedures. This is a typical example of "bureaucratic" Regulator.

The Third type of Regulators are known as "lazy" Regulators. These are Regulators who usually do not respond in a fast and agile way to any change in the international regulation or to the changing market conditions. Even the companies are waiting too long for Approvals or registration from these Regulators. This causes problems with companies, because there are Risky Industries which are internationally oriented (oil and petroleum, aviation, railway, etc.) and companies in these industries have a need to adapt fast to the changes. They are losing money if they do not adapt fast to the new market conditions and the "lazy" Regulators are a problem for them.

The reason for the existence of such behavior can be different, but it is not necessarily that these are bad Regulators. Maybe there is not enough staff or maybe changes in the industry do not happen very often or maybe there is not enough knowledge about the regulation, so they need time to understand it.

Whatever the reason is – it cannot be used as an excuse! The companies in the States with such Regulators do not feel appreciated, and they struggle to reach particular level of their business in the international market. With the lack of regulation or responses from the Regulator, these companies are doing their job, usually improvising, and improvisation is not safe at all. In addition, it costs them more money later when the regulation is passed.

The fourth type of Regulators are "over-regulated" Regulators. The regulation which these Regulators are producing is usually very detailed and almost every aspect in the area is the subject of some regulation. The problem is also that this regulation is full of legal phrases and legal complexity, so it is not easy to understand it. It may provide good rules for Oversight, control, and monitoring, but this type of "over-regulation" usually is "killing" innovation and the development of new products, new processes, or new

---

* Actually, after the *Deepwater Horizon* accident which happened in April 2010, this agency was accused of poor regulatory Oversight and the agency was split into three agencies.

risk control measures. If the company is dealing with overly detailed rules, then the company cannot build competition with other companies because all of them need to do the same: Abide by the same detailed regulation!

Usually, with such Regulators, the companies are subjects of additional costs which are coming from record-keeping and from reporting every event to the Regulator. These costs are reducing market competition and they are increasing prices for the company products and services.

In addition, these Regulators are usually very bureaucratic and the audit made by them could be very big burden for the company: It will last too long because too much documentation needs to be produced and to be audited.

The "over-regulation" is actually the response to a lack of self-confidence with these Regulators. They know that they can be the subject of investigation if something goes wrong and they think that "over-regulation" can protect them. "Over-regulated" Regulators are making their own job easier (they feel themselves to be on the safe side!), but the companies are struggling.

The real question is: Is safety improved by "over-regulation?" I have serious doubt about that!

In general, the main point is that "bad Regulators may destroy good companies."

And there are (of course!) "good" Regulators. They are rare (in my "humble" experience!), but there are some. These are Regulators which understand their position in the State administration and their influence on the business, so they are really trying to improve it. They understand that they are Masters and Slaves for the companies: Masters when there is something which must be done and Slaves when the company is asking for help. These Regulators are self-respecting, knowledgeable, experienced, and have partnership relations with the companies. They invest in their staff and they try to provide an excellent service to the State and to the companies. They are firm, but gentle. They are strong, but flexible. They understand "why" you are not-complying, but they do not hesitate to fine you.

It is pleasure to work with such a Regulator!

3.10 Who to "Blame" If There Is Non-Compliance?

Each Regulator is established as a State body with the intention to take care of the regulation and to protect the public interest. As such an entity, the Regulator has an obligation to oversee the performance of the companies and to take care they abide by the regulation. In addition, the Regulator has the power to punish the companies in the cases of wrong-doing. In the case of a Safety Regulator, the Regulator also has power to punish the companies or even to stop their operations.

So, there is a natural question which arises: Who to "blame" if there is non-compliance?

In Risky Industries, things are quite different than in other industries.

The science and the practice showed that safety cannot be maintained if there is no clear understanding between humans about the consequences of bad performance. This is the reason why the concept of Just Culture[*] was introduced in Safety Management, and this is the reason that I am recommending including in training for each procedure the possible consequences which can happen if the procedure is not followed.[†]

The power to "punish" the company is widespread in regulation, and it is very powerful tool. But (let me state it again!), in the scope of the context which is many times repeated in this book, the Regulator and the companies are partners in achieving safety. So I would not suggest using "punishment" as a tool to achieve safety.

Anyway, we are speaking for Risky Industry, where each bad performance or mistake/error can produce terrible consequences for humans, assets, and the environment. In these situations, sometimes "punishment" is only way to make the companies to abide by the regulation. The most effective "punishment" is to make the company to stop their operations. This will be terrible for the company and it must be used as last resort. This "punishment" for wrong or bad performance will have also social consequences, because plenty of people will lose their jobs and the unemployment rate will increase. That is the reason that this "punishment" must only be used if the Regulator is sure that a lot of chances were given to the company to restore the safe operations and the company failed to do that.

But let's go back and find the answer of the question: Who to "blame" if there is non-compliance?

Risky Industries are complex industries and usually things are not so easy to maintain. Anyway, in general, everything starts with the "fish." In the Balkan Peninsula, in the south of Europe, the people who are living there use one old adage which says that "The fish start to stink from the head." So, in accordance with this adage, the CEO (General Manager, Director, etc.) in the company is most important and most responsible person.

Most of the CEOs (General Managers, Directors, etc.) in the companies delegate the tasks to their employees, and they will defend themselves that the Safety Manager (or someone else!) was in charge of this, but this defense must not work. There are few reasons why this shall not be considered:

(a) The CEO (General Manager, Director, etc.) in the company may delegate the tasks (activities!), but he cannot delegate the responsibility! The responsibility always stays with him.

---

[*] See Section 2.19 (Just Culture) in previous chapter (Chapter 2)!
[†] See Section 9.2.6 (Auditing Procedures) in this book! (Chapter 9).

   (b) The CEO (General Manager, Director, etc.) of the company is the person who appoints the Safety Managers, so if something is not going as planned, it is his mistake for choosing the wrong person.

   (c) The CEO (General Manager, Director, etc.) of the company (regarding the SMS!), is responsible for regular Management Review of safety activities (SMS regulation requirement!), so he must be aware what is going on in the company.

So, regarding the "punishment," in cases when it is necessary, the Regulator should fine the company a considerable amount of money and at the same time, the CEO (General Manager, Director, etc.) of the company must be fined also by a fine expressed in money. Depending on the situation, even the Safety Manager (or other person in charge of the SMS!) should be punished by fine in money. What is wrong is to "punish" only the Safety Manager (or other person in charge with SMS!), because I have seen situations where good Safety Managers were limited in their activities by the wrong understanding of their CEOs (General Managers, Directors, etc.).

Do not look at me: I already told you that misunderstanding of safety is most critical with the Top Managers (CEOs, General Managers, Directors, etc.)!

Companies in each industry are economic entities. It means that they are working for profit. So, the fine in money could trigger some actions, but do not forget that companies are managed by humans and human reactions are very much unpredictable.

One of the fines which can be used for "punishment" of bad performance, and which I think should be used often, is revoking the Post-holder Certificate of the CEO or the Safety Manager. Of course, this should be done after considerable analysis is conducted by the Regulator. What is the point in having a person in charge who does not take care to follow the regulation? The Regulator, in such cases, must have in mind that: When the football team is losing games, the owner changes manager, not the players...

So, the two general recommendations regarding the "punishment" are:

   (a) The Regulators must use "punishment" as last resort!

   (b) If there is a situation where "punishment" must be activated, the Regulator must do it without demur!

4

Types of Audits

4.1 Introduction

Roughly speaking, there are two types of audits of Management Systems: Internal and External.

If you check ISO 19011:2018* document, on p.vi there is a table (which is presented in this book as Table 4.1), explaining the categorization of audits of Management Systems:

The main point is that the scope and subject of the audit will determine which type of audit will be conducted.

Anyway, the situation is not so simple: Whichever of the audits is used from the table, it can be classified as Documentation Audit and as On-Site Audit. Both of these audits will be explained in detail later.

4.2 Internal Audits

Internal Audit is the audit which company executes to check its performance. There are specially trained employees for the Internal Audit, and they check the functioning of the Safety Management System (SMS) implemented in the company. Usually, the Internal Audit is done once or twice per year and the results (findings) are given to Top Management and/or to the Safety Manager. They analyze the findings and they plan what actions (if necessary!) should be implemented.

There is a regulatory requirement for such an audit in each Management System† (not only for the SMS!) and I do believe that there is more damage than benefit with these audits.

---

* The name of this international standard is: Guidelines for auditing management systems.
† The Internal Audit is a requirement also for QMS, EMS, FMS, etc.

TABLE 4.1

Categorization of Audits in Accordance with ISO 19001:2018

First Party Audit	Second Party Audit	Third Party Audit
Internal Audit	External Provider Audit	Certification or Accreditation Audit
	Other External Interested Party Audit	Statutory, regulatory, and similar audit

The bad point with the Internal Audit is that it is done by the Internal Auditors who are employees of the Safety Department and this is wrong. The Safety Department is the department which deals with the maintenance and functioning of the SMS and the employees there are tasked to take care of the SMS on a daily basis. But the assumption hidden in the regulatory requirement for Internal Audit is that (obviously!) their job has nothing to do with monitoring, controlling, and improving the SMS on daily basis. Having this in mind, there is a simple and logical question: If the employees in the Safety Department monitor, control, and try to improve their SMS every day, why do they need an Internal Audit? Don't they gather information each day to assure themselves, the company's Top Management, the Regulators, and the public that they are safe?

Someone will say: They are checking on a regular basis how the company is following the SMS procedures, SMS rules, etc. But I had chances to be part of many Internal Audits in different areas and I could notice only a burden for employees, and I did not have any evidence that Internal Audits improved safety (or quality, or environment, etc.).

Having a Safety Department in the company is a necessity if the company is big. For small companies, the appointment of a Safety Manager is enough. And this is the reason that I do not like it when Internal Audits are done by the employees which are taking care of the SMS: If they do the audit, it means that they are controlling themselves. In such cases there is no impartiality of the audit and I have doubts about the objectivity of the findings!

Axiom 6: For an audit to be successful and to be able to improve a Management System, it must be impartial and independent! The impartiality and independence of audits are provided by auditors!

If the company would like to have an impartial Internal Audit, then it must not be done by employees in Safety Department. One solution is: The Internal Audit can be done by the employees in another department. So, if the company decides to engage employees from other departments to conduct the Internal Audit, they will produce another problem: The company must provide training in Internal Auditing for them. Let's also mention that this is an additional burden for these employees. They actually once (or twice) in the year "switch" their "brains" to an activity called Internal (Safety) Audit, and this "switching" sometimes tends to be too formal and too bureaucratic. If the Internal Audit is formal or bureaucratic, then it is of no use.

But there is more!

Even if the Internal Audit is done by employees in Safety Department, usually it is done in the wrong way. I had a chance to be in a company which was pretty much dedicated to achieving compliance with the safety regulation. They even had employed a guy as Compliance Manager with the task of monitoring and controlling compliance. He was taking care of the Internal Audits, but in the wrong way.

There were First Party Audits and there were Second Party Audits in the company. The wrongdoing was that both types of audits were Internal Audits, done by the employees in the company. The First Party Audits were conducted for any department in the company and they were done by the dedicated employee in these departments. So, the audit undertaken by them means that they actually control themselves. The Second Party Audits in the company were conducted by the Compliance Manager himself or by his assistant. Having in mind that both of them are part of the company, I would ask where the impartiality has gone?

But that what was worst was that both audits were done with regard to a regulation which is wrong! I know, I know, you are shocked: Why it is wrong?

Let's explain this in more detail:

When the company produced the SMS, it was produced with the intention of providing the rules and procedures which need to be implemented in the company to satisfy the regulatory requirements for safe operations. All these rules and procedures are explained in the SMS Manual. The SMS Manual is an official document of the company which explains how the company is monitoring, controlling, maintaining, and improving the implemented SMS. This document explains (in detail) to the employees what needs to be done to provide successful and sustainable safety operations. Having in mind that the company's SMS is already approved by the Regulator, the Internal Audit should not cover regulation! It is already satisfied by the SMS produced and implemented by the company!

What is important is that the Internal Audit should take care of the implementation and the maintenance of the SMS in the company. The Internal Auditors should check that the employees are familiar with the Safety Policy, with the SMS Manual and the rules/procedures inside, are they trained for safety procedures, for reporting procedures, for emergency actions, do they use all these procedures, do they keep proper records in the proper places, do they analyze safety performance, etc. With simple words: Whatever is in the company's SMS should be subject of the Internal Audits! So, the Check Lists* (CLs) for the Internal Audit should be prepared with regard to the SMS, not to the regulation. Maybe you are shocked, but the Internal Audit is not about regulation, it is about successful implementation and maintenance of the SMS in the company.

---

* See Chapter 6 (Check Lists (CLs)) in this book!

Imagine companies with "personality" Class 4 and 5* conducting an Internal Audit and this audit showed that there is some non-compliance. The point is that this non-compliance was present during the last Regular Audit conducted by the Regulator, but the Regulator's auditors did not notice them, and there was nothing about them in the Final Report from the Regular Audit. I would not say that the Regulator was bad and they did not do their job. Maybe these areas were simply not part of that audit or maybe the emphasis was put on something else, but the fact is that Regulator did not get findings about this non-compliance. And now, the Internal Audit findings report the non-compliance items, which I must say are minor (Level 2 findings).† To solve these non-compliance items, there is a need for additional resources and their solution will be costly and time-consuming. There is a logical question here:

What will the CEO (Director, General Manager, etc.) of this company do?

Do not forget: We are speaking about companies of "personality" Class 4
    and 5.

The only logical answer is: The CEO (Director, General Manager, etc.) of this company will simply postpone fixing (even maybe neglect!) these findings! Why does he need to put effort (resources!) into something which is minor and was not even detected by the Regulator? So, maybe he will wait few months before the next Regular Audit by the Regulator and then he will order corrective actions regarding these findings. Or maybe, he will wait for the Regular Audit and see what will be the findings there.

You may be shocked, but read Section 5.2 and you will see the logic behind these actions for each "personality" Class.

In my humble opinion, the Internal Audits could be deleted from the Safety Regulation, and actually, they should be deleted as a regulatory requirement from any regulation about any Management System!

I am not saying that Internal Audit will not bring benefit to the companies. I am just saying that the benefit is too small and is limited to the small number of companies (those from "personality" Classes 1 and 2 (maybe 3). Only these companies are willing to put efforts into something which is overestimated.

As I mentioned before, the employees of the Safety Department in each company must take care of the SMS on daily basis using daily provided information regarding performance of the implemented SMS. They should provide procedures for monitoring the performance of the SMS on a daily basis. For them "the audit" is conducted every day and it lasts forever!

---

* Classes regarding "personality" of the companies are explained in Section 5.2 in the next
    chapter (Chapter 5).
† Different Levels of findings are explained in Section 7.4 (Type of Findings) in Chapter 7.

If you are shocked (again!?), let me ask you a few questions: Does the company audit its daily operations (production processes)? Is there a yearly plan for an Internal (Production!) Audit for daily operations? Do not forget that we are speaking about a Production Management System* which conducts daily operations (production!).

They are doing daily Oversight (monitoring), but they are not auditing daily operation (production processes!). So, if there is problem in the product (or service offered), the Quality Control Department (Unit) will raise an alarm and the Quality Assessment Department (Unit) will investigate what the problem is and how it can be solved. This is done on a daily basis, but there is no audit! So, the same thing should be implemented for the Safety Department: They should monitor SMS performance on a daily basis and whatever happens (if anything happens!), they will fix it immediately! Of course, that lessons-learned process is implemented for the future activities!

Anyway, each company may do whatever they think it will bring them benefit and what is not against the regulation. But let's be honest: The regulation is requesting Internal Audits and there are few reasons for that.

The first reason is that Safety, Quality, Environmental, and all other Management Systems were upgraded to the ordinary Production Management System (which lasted for ages!) in the last 70 years. From the dawn of humanity, Humans had a production system which did not change too much during the centuries. So, the Organizational Era (see Section 5.1 (Introduction) in the Chapter 5!) brought new rules to provide better Quality, Safety, Environmental Protection, etc. Having in mind that these new rules were not embedded into Human habits, there was a need to control them, and Internal Audit was the tool which was used. But, after 70 years of use, I do believe it is time to change our approach to Safety (Quality, Environmental, and all other Management Systems), and to change Internal Audit to the daily monitoring (Oversight!) of safety (quality, environmental, etc.) performance.

The second reason is that the (new!) regulatory requirements were a burden[†] for the companies, because they needed to engage people who will deal with monitoring, control, and maintenance of all these Management Systems. And Regulators made a compromise by providing a regulation which asks for audits done by company itself. The costs for independent external audits are reduced by allowing the companies to do the audits by themselves. So, this was more a political and economical solution than something which needs to improve Safety (Quality, Environmental, and all other Management Systems).

There is one important thing which is part of our reality and which is not known by many businessmen: All the world best watchmakers (Rolex,

---

* Someone will say: There is no such system, because there is no such official (worldwide!) standard for requirements for such a system. But have in mind that each company has a Production Manager (or whatever it is called!) and where there is a Manager, there is a Management System.

† See Section 3.2 (How to Pass a Regulation) in Chapter 3 of this book.

Omega, Longines, Hublot, etc.) have not implemented any formal Quality Management System!

Can you notice the paradox?

These are companies which provide products of impeccable quality,* but they do not have a formal QMS and they do not conduct Internal Audits! Simply, their overall Production Management System is a Quality Management System. And it works! Perfectly works!

So, if someone would like to tell me that there are benefits from Internal Audits in any type of Management System: Sorry, I agree that there is a small benefit, but the Internal Audit is useless in most of the companies!

4.3 External Audits

There are two types of external audits: Second Party Audits and Third Party Audits.

4.3.1 Second Party Audits

Second Party Audit is the audit conducted by the (primary) company and executed over the other (secondary) companies which have a contract with this company for providing raw materials, parts, or services. There are a few examples of such cases:

(a) Any other company which is supplier of raw materials used in production process for the primary company.

(b) Air Traffic Control (ATC) has a contract with telecommunication provider for communication services.

(c) Airport has a contract with an electrical company for providing power supply.

(d) Nuclear power plant has a contract with another company to take care of already used uranium/plutonium.

(e) Pharmaceutical company has a contract with another company for providing chemicals used for medicaments production.

(f) Hospital have contract with a company which will gather and destroy the medical waste in the proper way.

The main point is that all these (primary) companies have a responsibility to take care of the safety of the products delivered or services offered to them, and it does not matter that some of the products (materials or devices) and

---

* A Longines commercial states that their watches provide accuracy of ±5 seconds per year!

services are undertaken by other companies. It means that they are obliged, during contract preparation, to put in a clause that they will control the performance of the contracted company and they will conduct a Second Party Audit to get assurance about the capability of the contracted company to contribute to the quality and safety of the "primary" company.

Another case for the need for a Second Party Audit is when the company is actually organized as a Holding Corporation or when there is any type of partnership between two companies. In such a case, a Second Party Audit is done from the "mother" company (HQ, Holding Corporation) to the companies which are "daughters" of this "mother" company. The trained auditors from the "mother" company (or maybe engaged consultant companies) will provide audits on the "daughter" companies and they will prepare reports on the performance of these companies. Very often, especially in the cases of engaged consultant companies, they will propose measures to improve the performance of audited companies.

Impartiality and quality of these audits is usually very high, but it comes at a cost!

4.3.2 Third Party Audits

The Third Party Audit is the best type of audit because it is independent from the company and it can be very useful to discover hidden organizational risks inside the company. Every Internal Audit is the object of subjectivism because it is conducted by the company itself and in general, it is not good when "we control ourselves." Maybe the Internal Audit will produce some findings that some of the procedures are not followed, but big errors (especially systematic errors!) in implementing or maintaining the management system can be found only by External Audits.

Usually, all the categories of Third Party Audits are done by Certification Bodies or Regulators, but sometimes good companies organize such Third Party Audits by themselves to get realistic picture what is going on with their companies. Companies may sign a contract with any consultant company competent to auditing a particular Management System and they can provide Third Party Audits at regular intervals. This type of outsourcing is very common for financial audits, but other areas could be subject of this outsourcing also. During these Third Party Audits, the engaged company is also providing consultancy, so the company can really have enough data to improve itself.

Anyway, it is not unusual if even the State Regulator signs a contract with some auditing company for audits, but I strongly do not recommend this type of outsourcing. There are a few reasons for such a recommendation:

The first reason is that the Regulator will transfer only the activity of the audits, but he will still be responsible for Oversight. So, what is the point in being responsible for an activity which has been delegated to another company?

The second reason is that the Regulator will lose the respect of the companies which he is obliged to oversee. This step is sign of weakness and nonseriousness and this is, in general, not good for the Regulator. This step also will send a message to the companies that the Regulator is pushing them to comply with regulations (and they spend a lot of money for that!), but he is not ready to equip and train their own employees who will oversee the compliance with regulation.

The third reason not to do it, is that after the audit, the Regulator will receive only the Audit Report from the engaged company, but he will not get clear information how the company behave in regard to its safety performance. It will happen because the Regulator was not present during the audit. The absence of the Regulator inside the company will not provide enough data for good oversight activities which are part of the Regulator's responsibilities.

However, there is a possibility for Regulators to provide different type of outsourcing regarding the audits. This outsourcing is based on establishing a pool of experts recognized by the Regulator to do oversight activities. This happen often with international regulatory organizations (bodies), especially with the agencies of United Nations. The Regulator in such a case does not engage a company, but he invites the independent experts to join the Audit Team (established by the Regulator) and the job is done by them on commercial basis (the experts are paid for that). Such a situation exists with ICAO* during their USOAP (Universal Safety Oversight Audit Program). The USOAP is a program which help ICAO to oversee the State Safety Program (SSP) regarding their fulfillment of the ICAO requirements for safety performance of the State's aviation systems. Each State is subject to this audit at least once in two to three years and the ICAO Team (established from ICAO pool of experts) is in charge of that. Employing such a big number of auditors which will take care for audits of all ICAO Member States is too expensive, so establishing an Audit Team by outsourcing to individual and independent auditors is reasonable practice. I think that a similar process is used also with IAEA (International Atomic Energy Agency), but I could not find reliable data to confirm it.

4.3.2.1 Categories of Third Party Audits

Third Party Audits can be classified into five categories.

The first category is Approval (Certification) Audit[†] and it is done as part of the Approval process of the Regulator or for Certification purposes by a Certification Body. This is an activity which results from a combined

---

* ICAO stands for International Civil Aviation Organization (the UN's specialized agency for aviation).
† In the book, I will use term Approval Audit. Whatever is said about Approval Audit applies also for Certification Audit.

audit: First, the Documentation Audit is done and when it is satisfactory, the Regulator (Certification Body) provides the On-Site Audit.

Each Management System implemented in the company must be based on particular documentation* which explains the rules, responsibilities, processes, and procedures which govern the associated Management System, and it shall be documented. As part of the Documentation Audit, at the beginning of each audit, the Regulator requests the company to submit soft and/or hard copies of documentation which will be the subject of assessment by the Regulator. The documentation is used to familiarize the auditors with the Management System under audit. In addition, the submitted documentation must be thoroughly assessed to see does it provide enough proof that every paragraph of the Regulation is satisfied. When the auditors are familiar enough with the Management System and they are satisfied with the compliance of the documentation, then they will schedule the On-Site Audit. The aim of the On-Site Audit is to check that what is written in the SMS documentation has been implemented, monitored, controlled, and maintained in reality.

The Approval Audit† is most comprehensive and complex type of audit and this audit will be the primary interest of this book. As said above, it consists of the Documentation Audit and the On-Site Audit. Whatever is mentioned regarding the audits in general totally applies to this category of audit. Other categories of audit are usually simpler and for each of them, the Regulator will need fewer resources and less effort.

The second category is the Regular Audit. It is an audit which is conducted once or twice each year and it can also be called a Continuation Audit. The point with the Regular Audit is that it is intended to continually oversee the already implemented and approved (Certified) Management System. For the Management System, it is not enough just to be implemented. It must be continually monitored, controlled, maintained, and improved all the time!

The Regular Audit may consist of Documentation and On-Site Audits, but usually Documentation Audit is very "rudimentary." It means that if there is no change of company documentation (new process, new procedure, or some other change!), then only the On-Site Audit is conducted. Documentation of company's Management System has to be checked by auditors as a reminder of what was going on during the last year and what happened in between the two Regular Audits.

---

* The most important documentation for each management system is the Manual! In addition, there could be other documentation and this is company-dependent. Some of the companies may use a different method of organization of their Management System (pay attention to Axiom 3!).

† In the literature, you can find this audit under the name Verification Audit because this is simply process of verification by the Regulator that the Management System of the company complies with the requirements. Please note that the name of this type of audit may vary, but whatever is mentioned in this book is still valid.

The third category is the Follow-Up Audit. If during the Approval or Regular Audit, there are some kinds of non-compliance, the company will be tasked to solve the problem (rectify non-compliance!) in the particular period of time. This Follow-Up Audit is done by auditors to check only the rectification of the non-compliance found in the previous audit.

The fourth category is the Exceptional Audit. This is an audit which is organized if there is information (from the company or from some other source) that there is something wrong with the company's safety performance. This information (even it is rumor) can create doubt that the things are going on in the proper way, or there is just a need for an additional check of the company. This is a common audit in Risky Industries. For those industries, there is a regulatory requirement to report any event which could (have) endanger(ed) Humans, assets, or the environment. So, the Exceptional Audit can be triggered by any such information and it is the obligation of every Regulator to check what is going on. Depending on the gravity of the situation, in such cases the auditor (Audit Team) may be even accompanied by the police.

The fifth category is the Special Audit, which is very similar in organization to the Exceptional Audit. It is conducted when there are situations where the Regulator has requested from the companies the implementation of some measure (or regulation) which will improve the overall or particular safety performance. Usually a particular time period is given to implement the measure (or new regulation), and after that Regulator organizes a Special Audit to check the implementation.

Later in the book, when I explain all audit activities, I will provide more details about Approval and Regular Audits, and this information will help auditors in conducting all other audits.

4.4 Frequency of Audits

There are many types of audits and the use of all of them are explained in previous sections. But there is a normal question: How often should each of the audits be conducted?

There is no right answer to this question!

It differs according to the industry and the type of audit. Regarding the frequency (how often they should be conducted) of audits, we can roughly categorize them into two groups: Regular and Irregular.

Looking of the names of different categories of audits, you can notice that there is one with the name Regular Audit. This is the only audit which must be done at regular time intervals, which, I must say, are not so "regular." Speaking about regularity, it means that it will be done each year, but there is possibility that one year the audit will be done in April, next year in

May, third year in September, and so on. All this depends on how busy the Regulator is and how the companies are performing.

In my humble opinion, I can say that, at least once in the year, each company should be audited.

If the "personality" of the company* is Class 1 or Class 2 and the results of previous audits showed no findings, there is nothing wrong if Regulator decide to extend the Regular Audit to once in 18 months or once in two years. It does not mean that for 18 months (two years) the company will not be overseen. Please remind yourself that the Regulator has the legal responsibility to oversee the companies and audit is just one tool for Oversight. He may use some other tools for Oversight in between the audits, or he can execute other types of audits (Exceptional or Special Audits) instead Regular Audits.

The frequency of the Regular Audits for the particular company must be established by considering:

(a) Size of company

(b) Volume of production (products or services)

(c) Availability of information gathered through other oversight tools

(d) Sensitivity and frequency of received information

(e) "Personality" of the company

(f) History of the events

(g) History of findings

Nevertheless, if the frequency of the audits is not part of the regulation, it should be! The Regulator may (and should) decide what and how often it has to be done. But (again!), I am proposing a balanced approach which will give the opportunity to the Regulator to use, but not to abuse, these rules. There is a need for a yearly schedule of audits to be published and to be provided to the companies. More about this is given in Section 8.2 (Schedule of Audits).

All other types of audits are triggered by a particular "safety event"† and there is no need for regularity in conducting them. But if such a need for "regularity" shows up (the Regulator very often receives information which requires an Exceptional Audit), it is a clear sign that things are not going well with the particular company and it needs to be checked with a more thorough audit (a Regular or Approval one).

Imagine a situation where the Regulator must conduct three Exceptional Audits in one company for seven months in a row. An Exceptional Audit is an audit which is triggered by information that something is going wrong and the Regulator in a situation when there is such information for the third

---

* "Personality" of the companies is explained in Section 5.2 in the next chapter (Chapter 5).
† For the purpose of this book, the term "safety event" is used with the meaning: Any event which will increase the belief that the safety performance of the company is experiencing problems.

time must remind himself: Was the information on the first occasions valuable? Was the information just "fake news?" Were there any findings? How serious were these findings?

If the previous two times the information was "fake news," there is still a need for an Exceptional Audit. It is good if the Regulator checks every time there is such information given, but again, it is subject to the human availability for such a purpose. Anyway, if there were some findings the previous two times, then the Regulator must schedule a Regular Audit as soon as possible: Three "problems" in seven months is a clear sign that the company is struggling to maintain its activities and it needs comprehensive "look" by the Regulator.

5

Companies and Audits

5.1 Introduction

You may think whatever you like, but you cannot run away from the fact that in your daily activities, you are implementing some level of trust to people, assets, premises, things, etc. And this also applies to the audit: The auditor executes his audit based on the level of trust which he can associate with the audited company. That is the reason for this chapter: You must have an understanding of different classes of companies to establish particular level of trust with them.

But there is also another reason for this chapter:

Following the history of the development of Functional Safety, it can be noticed that during previous periods, different things in industry were the subject of audit attentions. At the beginning (approximately from the 1920s to 1950s), there was the Technological Era where the development of technology was the main contributor to achieving safety in Risky Industries. This is a time when critical technical systems were doubled (two of the same systems in redundancy) and associated with monitors which would automatically change over to another one if one were faulty. It brought some improvement of reliability, but this was a reactive approach and the investigation of causes for accidents was just starting.

The next period happened after the 1950s and lasted approximately to the 1990s. It was known as the Human Era. The previous period's investigation of accidents showed that 80% of all accidents were caused by Human mistake/error. It does not mean that the Technological Era brought good equipment that never experienced faults. Reliability and integrity of equipment increased very much in the Technological Era, but equipment continued to fail. In such a situation, there were procedures which were written with explanations of what to do when equipment experienced faults. It was noticed that even in such cases (when there is procedure to follow), humans fail to follow the procedure and to provide a safe outcome in situations where the equipment is faulty. These were the first beginnings of Human Factors as contributing to unintentional human mistakes/errors.

Data gathered and analyses regarding Human behavior in the incidents and accidents showed that there are factors which are the reason for wrong human performance expressed by unintentional mistakes/errors. This contributed to the development of Human Factors as part of a psychological science. Later, it was noticed that working atmosphere in organizations (companies!) contribute very much to these human unintentional errors, and so the Organizational Era was born at the beginning of this century. This was actually a time where stronger and systematic regulatory requirements, especially for the implementation of Safety Management Systems (SMSs) were introduced. Further research showed that "Bad companies may destroy good people," so maintenance of Human Factors and Management Commitment became a requirement of each SMS.

That is the reason that each auditor in each audit must understand the big picture about the company which can be presented as the "personality" of the company.

5.2 "Personality" of the Company

As it has been said at the beginning of previous paragraph, to provide sustainable safety in reality, there is a need for a particular amount of trust which must be built between the Regulator and the companies. The trust of the Regulator towards the company is gained first through the Documentation Audit.

A good auditor will carefully read documentation submitted by the company (regarding the implemented SMS), and he will gather understanding about the level of knowledge, skills, experience, and attitude of the company to safety. For such purpose, the auditor may also consult the Regulator's archives (the history of previous audits!), reports, or events (if there are any in the Regulator's archive) regarding the company's performance. This history is important because it will help to determine (approximately) the present situation with the company.

There are different classifications of the companies regarding their attitude to safety, and one example for such a classification of them is presented in Table 5.1. The auditor must understand the different types of "personalities" of the companies, just to know what he can expect during the audit.

The "Personality" of the company can be defined as a combination of attitudes, style, and character of doing business of individual companies. Different classes of "personalities" are presented in Table 5.1.

Regulators are doing Oversight of the companies, and to be successful in these oversight activities, it is important for auditors to establish the "personality" of the company. Audit is just part of the oversight activities, but it is an activity where the Regulator staff is in direct contact with the employees of the company. Even in the same industry, different companies are using different equipment and they employ different humans with

TABLE 5.1

"Personalities" of Companies

No.	Class	"Personality"	How to Deal with Such a Company?
1	Generative	This is a company which has an excellent attitude to safety. All factors affecting safety are analyzed, and elimination or mitigation of the risks are implemented. Human Factors affecting safety are controlled by the working atmosphere which is excellent. This is a company which appreciates open minds and good ideas and they show an excellent level of Just Culture implementation. Spreading of safety information is encouraged, not only within the company, but also with the outside world.	It is a pleasure to work with such a company! They respond quickly to each request. They are open and sincere, and this is based on their confidence regarding the efforts and effects which they put in their everyday operations to maintain safety. They fulfill all obligations and promises in timely manner. The audit should be "relaxed" and it can be finished without any problems.
2	Proactive	They deal with safety by the book! Whatever is requested from them is done in accordance with the regulation, guidance material, or experiences of others. They have resources for statistical and probabilistic analyses and they are good at Risk Management. Although they are proactive, sometimes they are known to be bureaucratic and the courage for new ideas is missing. Just Culture is present.	It is good to work with such a company and usually they are covering all regulatory requirements. But having in mind that they are dealing "by the book," implementation of the system could not go so well. Sometimes the bureaucracy of these companies is known to be disputable: Everything is in the documentation, but not everything is functioning smoothly in reality. The audit should be a normal one, which means: No big surprises should be expected.
3	Calculative	These are companies which only comply bureaucratically with the regulation. Their decision is made more on calculation and economy than on safety considerations. Sometimes they undertake Risks to do something which can bring economic benefits, but they have a lack of understanding of the causes and consequences of the actions undertaken. Level of Just Culture is questionable: It applies only if it does not endanger economic benefit.	Level of bureaucracy here is bigger than that with the previous categories of companies. Lack of proper understanding of SMS is present. These companies have a Safety Manager and this is usually a very experienced guy who worked in the industry. But he is missing knowledge about the methods regarding Hazard Identification and Risk Assessment. They provide all documentation, but it is poorly effective. Usually procedures are complicated and they do not provide effectiveness and efficiency. You may recognize such a company during a Documentation Audit.

(Continued)

TABLE 5.1 (CONTINUED)

"Personalities" of Companies

No.	Class	"Personality"	How to Deal with Such a Company?
4	Reactive	They take care of improvement only if something happens. In general, they are satisfied by themselves and do not expect bad things to happen to them. Not aware of Just Culture and improvement is triggered based on economic benefits. Safety is questionable and it is only bureaucratically satisfied, with a lot of missing points.	Usually with these companies, Hazard Identification and Risk Calculation lists are missing. They do not think in advance (reactive approach to safety!) and there are no Preventive Procedures, or if they exist, they are poor. Even the records of events are not so good (something is always missing!).
5	Pathological	These are companies which do not take care of safety. They are based on hiding bad events and cover up activities of incidents and accidents. Their safety performance is on the edge of criminal intent. If something happens, somebody else is guilty! This is a company where a lot of incidents (or even accidents) happened in the past.	These are highly manipulative companies. They usually have implemented a formal SMS, but most of the documentation (especially records) is fabricated. It is pretty much frustrating to deal with such companies. They very often lie to auditors and the auditor must be careful because these companies will try to discredit the auditor if there is non-compliance (and usually there are many items of non-compliance!)

different levels of education. These humans are connected with equipment by different procedures. In different countries, there are different cultures and religions, so it is important to understand that every company is different in its "personality."

Axiom 7: Establishing the "personality" of the company will help you with the strategy and tactics to organize and conduct the audit!

Pay attention to the word "help" in Axiom 7! It means that auditor can use "personalities" of the companies just as help which can provide particular expectations for him regarding the audit. Whatever the "personality" is, it may not be used as a judgment! It means that, for the things which are not covered by regulation, the attitude and actions of the auditor during the audit cannot be triggered by his expectations regarding the company's "personality." Let's say, for the companies with Class 1 and 2, the auditor can be more confident that the company will be honest and sincere during the audits. This is not always the case with the company from other classes. In companies from classes 3, 4, and 5, the auditor must be more focused and he needs to put in more effort to establish the real situation.

5.3 Understanding the Audit of the Company

The auditor, in addition, to be successful in his job, must understand the answers to two very important general questions regarding audits in Risky Industries:

(1) What is the purpose of the audit?

(2) How are the companies reacting to audits?

I will try to provide answers on these questions in the next two sections.

5.3.1 What Is the Purpose of the Audit?

All companies are driven by profit! They are economic entities which produce products or offer services and they sell these products and services on the market. Very often, these companies, intentionally or unintentionally, neglect their obligations and duties to provide safe products or safe services and "bad things happen to good people." In any such case (intentional or unintentional), if there is damage to the assets or environment, or harm (death) to people, there is a need for an investigation on the level of the State. The State, through the Government, is responsible for providing a safe life for their citizens and protection from losing their assets. This is usually written in their constitutions or some other primary laws.

But this is a reactive approach to safety: Something bad happens and the State will react with legal means. The new approach in the safety area is: Think in advance! This is an approach where each company must establish a particular Safety Management System where the company, proactively, will deal with safety. It means the safety staff there will identify hazards in advance, calculate the risks, assess the acceptability of the risk and implement some measures to eliminate and/or mitigate the risk. In addition, they must spread the information regarding safety events to others with the intention of warning them and to help them, because these safety events may happen to them also.

To protect the public (citizens!) from the consequences of bad behavior which can produce harm or damage to assets and to nature and/or death of humans, each State has established a particular legal system which deals with all these things. Having in mind that economy and trade is globalized today, there are plenty of international organizations which are producing international rules for providing safety regulation for Risky Industries.

Some of them are international, belonging to the Organization of United Nations (e.g., International Civil Aviation Agency (ICAO), International Atomic Energy Agency (IAEA), etc.), some of them are established as State agencies in countries all around the world. Usually States voluntarily join international organizations. If they join, they are obliged to implement the organizations' regulation in their legislation system. In addition, the States have responsibilities to oversee* the implementation of these regulation by the companies on their territory.

Disasters happen on a large scale and it can affect a few countries or a few continents, so the existence of such international organizations and the States' membership in them is a normal situation, because crisis responses must be coordinated internationally.

The already stated main point is that all these organizations are providing rules and procedures which will be used by the States' Regulators and companies. But the most important is that they understand that "We need two for Love!"

It means that these organizations promote partnership (I would say "Love") between Regulators and companies. Partnership is quite different to the Master–Slave interactions which is very often promoted by States' Regulators! The partnership is based on a clear understanding of the responsibilities and obligations regarding what needs to be done by Regulators and what needs to be done by companies. Of course, Regulators have the power to sanction companies, but this power must only be used if necessary and must not be abused! The Regulators are not police, and they do not do audits just because they have the power to do them! They do audits as part of their oversight activities because this is part of their job.

Whatever the class of company is (regarding the Table 5.1), the auditor conducting the audit as part of the regulatory oversight activities must not behave as an arrogant person who has the power to destroy the company. The auditor must do the job in accordance with the partnership obligation, established by the common understanding. It means that the auditor must listen the arguments of the companies and must understand them. He must not use his opinion as a judgment, and he will act in accordance with the regulation only if there is objective evidence that something is wrong. This is especially important for Class 1 and Class 2 companies where the common understanding and the common trust is biggest.

Axiom 8: Safety Regulators and companies are not connected in a Master– Slave relationship! They are partners in providing safety!

The simple example of how this partnership between the Regulator and companies must be is the relationship of the older brother and younger brother in a good family. The proper older brother will take care of his younger brother.

---

* Audit is just one of the tools used to do Oversight.

He will monitor the younger brother's activities and behavior and he will protect him. The older brother will behave in such a way as to establish a good example for his younger brother of how he needs to behave. But if there is need, the older brother will not hesitate to rebuke his younger brother. Such behavior of an older brother can only provoke respect in the younger brother. But if the younger brother listens to the older brother because he is afraid of him, things are not good.

This partnership is something which is very much missing in today's relations between the any Regulator and any company (in any industry).

The reason that the auditor must keep this attitude is because the style of auditing speaks about the Regulator's attitude to safety and it must be impeccable! In general, you may not ask for respect if you do not provide reasons for this respect. Respect does not come with the title! It must be earned!

Partnership must take care that the obligation of the company is to provide the data requested for audits and the obligation of the auditor is to check these data looking for objective evidence if it is something good or something wrong. Later, I will emphasize this part in more detail.

5.3.2 How the Companies Are Reacting to Audits

Companies are not happy with audits at all!

The unhappiness of companies with audits is reinforced by the possibility of being fined if the company does not comply with the regulation, so each audit for them is a chance to lose money. And companies hate to lose money!

There are a few reasons for such an unhappy attitude of the companies:

The first reason is that, in reality, most of the companies do not have a good understanding regarding safety. Mostly companies are employing Safety Managers because there is a regulatory requirement for that. The CEO (the General Manager) is not familiar with safety principles and he does not take into consideration the requirements for Top Management's dedication to safety. In his humble opinion, Safety is the responsibility of the Safety Manager, and his job is finished once he appoints such a person! And this is something to which the auditor must pay attention during the audits: Check the Management Commitment to Safety! Check is it realistic or just bureaucratic.

Unfortunately, and most often, even the Safety Manager is not a guy who has considerable training in the area of safety. Very often, Safety Managers are appointed persons who are experienced employees and the safety system in the company is based on their experience (if something happens, they will know how to deal with it!) Maybe you can be shocked by this statement, but go to any job portal on internet and try to find a job opportunity for a Safety Manager. If you understand safety, you will be shocked by the requirements which a requested person for Safety Manager must satisfy. I am not speaking in general, for all industries. I am speaking for Risky Industries, especially for aviation. This is a common situation in most of the industries, but it happens, especially to Class 4 and Class 5 companies from Table 5.1.

The second reason for reluctance towards audits is the natural wish of humans not to be audited (controlled!). All of us like our independence, and we oppose in general any way of being controlled. But the State does not control companies because the State enjoys control. The State controls companies to protect the public interest and to provide the rule of law. This is often forgotten by the companies and unfortunately, it is forgotten even by the State (by the Regulators!)

The third reason is that companies do not like audits because they are influencing their daily operations. Whichever type of audit it is, they must dedicate a particular number of their employees to it, taking them out of daily operations to help auditors with the audit. I remember that one UK aviation auditor told me a story of an audit on a small British airport when the CEO of the airport asked the auditor:

> "There are 116 days in this year, when the airport was visited by some kind of regulatory auditor. How do you expect to do my job if I need to spend 116 days out of 365 with auditors?"

Having in mind above-mentioned statement, each auditor must thoroughly prepare for the audit and must dedicate all his efforts to provide the audit is as short as is possible. In addition, the audit must not disturb daily operations (as much as is possible). That is not easy, and that is the reason that the book is named *The Art of Safety Auditing*. Safety Auditing, by itself, is a creative process where the auditor must have the capability of a good understanding of the situation, particular professional knowledge, Human interactions' skills, and adequate personality, so auditing really can be defined as Art.

The auditor must understand that an audit, although required, is a burden for the companies. That is the reason that a regulatory audit must be properly organized and planned regarding the members of the Audit Team and the schedule of activities. The Regulator needs to propose such a timing to company, so the audit produces as little disruption as possible of company activities. The date when the Audit Team will come to visit the company must be coordinated with them, and it is strongly recommended to inform each company at the end of the year of the dates of audits in the next year. Some Regulators provide a list with the planned audit dates on their website, which is also OK. If the company is not happy with the audit dates, Regulator must understand why they are not happy. If the reason is operational, then you may take into consideration this dissatisfaction and change the audit dates.

But if a company from Class 4 and 5 from Table 5.1 disagrees with the audit dates, this is often not because audit will influence their operations or planned activities. These companies do not comment on the yearly Schedule of Audits when it is published or sent to them. They will react to the letter from the Regulator when the Team Leader will state precisely on which date the On-Site Audit will be conducted. They will try to postpone it, and in such a situation, it is good to postpone the audit, if there is a chance for this to be done. In most cases, they are panicking because they have forgotten the Schedule of Audits. They need more time to fix problems or deficiencies of their SMS and that is good! Postponing the audit will give them a chance to fix the problems and it will make their SMS better, at least for a short period of time. And this is the target for every Regulator: To improve the safety performance of the companies in their own State! But be careful: It could happen very often and I do not know what to advise in such a situation... Anyway, it is prudent, if there is such a company, that they need to be informed regarding upcoming audits at least two months in advance. It is maybe more of a burden for the Regulator, but it can increase the audit's integrity.

Axiom 9: The audit must be well-prepared and organized by Regulators (in partnership with the companies!) to be effective and efficient in regard to resources and time!

6

Check Lists (CLs)

6.1 Introduction

In general, the Check Lists (CLs) are an aid (tool) which is used in any industry. They serve different purposes in different situations, but I will explain their use in this book only for the purpose of audits.

In general, CLs are used as an assurance or reminder that a job (operation, task, or activity) has been done or needs to be done. For example:

(a) Pilots use CLs to do all preparatory work for take-off or landing. This is the way to assure themselves that aircraft is ready for take-off or landing.

(b) Professional divers use similar check lists before diving; these CLs are simple and they are represented by questions which provide answer in the form Yes or No (Done, Not Done).

(c) Doctors fill CL forms before accepting patients in hospitals and/or before surgery. The questions there are different and in such cases, the patients are providing information which can contribute to their safety (allergies, other receiving medications, etc.) or to their healing (previous medications received, surgeries done, etc.).

(d) In software design, CLs are used for checking process compliance, code standardization and error proofing.

Audit CLs are different because they rely on regulation. In auditing, all regulatory requirements are placed in one or a few different CLs and the auditor uses it as a tool which will help him to check compliance with regulation.

As you can notice, the wording "Check Lists" is written in plural, because one Check List cannot serve for the audit. Usually there are a few areas to cover (each area for a different type of regulation) and there is usually need for a few auditors for each area. So, during audits, there is need for as many CLs as there are areas for auditing. For example, when Auditing Air Navigation Service Provider (ANSP), the Regulator needs to audit air traffic control (ATC), engineering services (CNS), meteorological services

(MET), aeronautical information services (AIS), and search and rescue services (SAR). Similarly, in a nuclear power plant, the Regulator needs to audit the nuclear reactor, steam turbine, generator, cooling system, etc. And all of these parts are subjects of a different regulation,* so they will need different auditors and different CLs.

Check Lists (CLs) in regulatory audits have to provide help to auditors on what to ask and where to look, and what type of evidence is needed, and there must be a place in the CLs to provide some comments. So, they are usually made as tables with columns and rows, there are small boxes for ticking inside, and there is space to provide additional remarks or comments in each row. CLs can be done on computer in Excel or in Word and may be printed or uploaded on a tablet for use during the audit inside the company's premises. I have a laptop and a good tablet also, but I prefer to have CLs on paper and use a pen and handwriting to fill it. The reason for that is that audit is done on-site, in companies or factories, where the convenience of paper is superior to laptops and tablets: If paper falls down, it will not be damaged, but laptops and tablets are prone to damage. If you do not believe this, try it…

CLs in auditing are something which needs to be done by each Regulator. They should be done for each area of auditing and must cover all regulation requirements. But do not exaggerate! Be careful how you prepare a CL: You do not need to put every regulatory requirement in it. A simple example is that there are some specifications regarding the equipment performance which needs to be satisfied. You do not need to write a particular row for every specification. You can just write "Specification of Equipment" in one row and there you will ask for the printed list of all specification performances, checking them later. Usually you may group the requirements on the basis of the expected answer or on the basis of the places (areas) where these requirements need to be checked.

Axiom 10: The auditor shall use the Check Lists during the audit! An audit without Check Lists is a "struggle for survival" with a very pessimistic forecast: The auditor does not know what to do and when and how to do it!

As a good behavior of the Regulator and as a gesture of a supportive partnership between Regulators and companies, Regulators should provide the CLs to the audited companies in advance. They can deliver them few weeks/ months before the audit or in my humble opinion, the best way is to publish all of them on the Regulator's website to be available not only to companies, but also to the public.

---

* Safety Regulation is the same (every aviation subject shall implement SMS), but the area of implementation is different, so each of them will have different procedures.

6.2 Two Types of Check Lists

In the audit process, there is a need to prepare two types of Audit CLs. The first type is used as a reminder regarding the targets of the audit and they are called Target Check Lists.* The second type is used to conduct the audit and they are called Audit Check Lists. In the following section, I will explain Target CLs and in the next one, I will explain Audit CLs.

In general, if I mention in the book only CLs, the meaning is: Audit Check Lists. When I would like to point to Target CLs, I will put word "Target" before the CLs acronym.

6.2.1 Target Check Lists

There are two types of audit targets which can be implemented into one CL: A High Level and a Low Level target. Both of them are expressed as items which are prepared as a list before the On-Site Audit, and they are actually used to remind the Team Leader (and auditors!) why the audit is undertaken (High Level targets) and how the High Level targets of the audit can be achieved (Low Level targets). Target CLs are produced by the Team Leader and the auditors in the Audit Team must be familiar with them, but during the audit, the Target Check Lists are used only by the Team Leader. When each auditor is trying to achieve Low Level targets during the audit, he is using only Audit CLs without any reference to Low Level targets.

There is a big difference between the Audit CLs and Target CLs, because the Audit CLs are the same for each type of audit (they do not change!), and the Target CLs are different for each type of audit. The Target CLs are different even for the same type of audit which is conducted in different time periods.

Table 6.1 is a simple example of a combination of High Level targets and Low Level targets given as one Target Check List for a Regular Audit. High Level targets are presented with bolded and italic letters in the rows (in the Table) and all other rows (with normal letters!) are Low Level targets. Please note that this table is just for the purpose of explaining High Level and Low Level targets which are integrated into one Target CL. And please have in mind that this is not a suggestion of how to do it. Every Team Leader can produce his own solution for these Check Lists. What is important regarding Target CLs is that they determine the direction of conducting the Documentation and On-Site Audits and keeping this direction is the obligation of the Team Leader.

---

* The experienced reader can notice that the Target Check Lists are strongly connected with the scope of audit and this is not wrong! Actually, the Target CLs are an operationalization of the scope of the audit (the scope of the audit needs to help achieve the items in the Target CLs). As such, the scope of the audit (through the Audit Schedule) should be presented to the company, and Target CLs are used by Team Leader.

TABLE 6.1

High Level and Low Level Target CLs for a Nuclear Power Plant with Four Reactors

High Level Target Check List for Regular Audit for 2016		
No. **High Level Targets**	**YES**	**NO**
1. Check situation with reactor 1		
1.1 Check records about radiation measurements		
1.2 Check records of outages of reactor		
1.3 Check records of cleaning a fluid system		
2. Check situation with reactor 3		
2.1 Check records about calibration of radiation measurements instruments		
2.2 Check records about concrete construction tests		
2.3 Check the records about Data Evaluation		
3. Check electricity distribution equipment in reactor 2		
3.1 Check the High Voltage Transformer(s) records of maintenance		
3.2 Check the records about high-voltage insulation measurements		

High Level Target Check List for Regular Audit for 2017		
No. **High Level Targets**	**YES**	**NO**
1. Check situation with reactor 2		
1.1 Check records about radiation measurements		
1.2 Check records of outages of reactor		
1.3 Check records about steel construction tests		
2. Check situation with generator 4		
2.1 Check records about preventive maintenance of mechanical parts		
2.2 Check records about temperature inspections		
2.3 Check records about calibration of radiation measurements instruments		
3. Check maintenance procedures in reactor 4		
3.1 Check preventive maintenance records		
3.2 Check corrective maintenance records		
3.3 Check Follow-Up actions		

In the cases of Follow-Up, Exceptional, or Special Audits, there is no need for Target Check Lists. All these audits always deal in advance with a specified activity and the target is usually a well-known one:

(a) For Follow-Up Audit, it is a check of fulfilling the actions to rectify non-compliance.

(b) For Exceptional Audit, it is a check of information gathered from official or nonofficial channels.

(c) For Special Audit, it is a check of implementation of new regulation or new measure.

Whatever you are thinking about Target CLs, please have in mind that for Approval and Regular Audits, these CLs are important and inevitable.

The Regular Audit is conducted usually once per year for each company, and every next such audit is the same type of audit. The difference comes from the fact that although it is the same type of audit, the same things are not checked in any two consecutive Regular Audits. It means that the High Level and Low Level targets in each consecutive audit are different. In accordance with Section 8.5 (Sampling) in Chapter 8 of this book, you cannot check everything during one Regular Audit, so the Team Leader is sampling different targets which need to be audited for each audit. Summarizing* the targets from a few successive audits, the Regulator can oversee the overall performance of the company over a few years.† Maybe it looks strange, but this is the way auditing functions and this the only reasonable (and practical!) way to do an audit in the shortest possible time.

As an example of Target CLs, I would like to present a specimen for Target CL (Table 6.1) for Regular Audit of nuclear power plant‡ with four reactors. Usually during each Documentation Audit, everything is checked (all documentation!), but during each On-Site Audit, only one or two reactors or generators, randomly chosen, are checked. In general, a different part of the company's implemented Safety Management System (SMS) is used for audits conducted over a few consecutive years to obtain a summarized (integrated) picture of how the company's SMS is performing.

Following Table 6.1, it can be noticed that, if the first year's (2016) Regular Audit's High Level targets are reactors 1 and 3, in the next year (2017), the audit High Level targets should be reactors 2 and 4. Looking at Table 6.1, you can notice that even other (Low Level) targets for these two successive audits are different. If you check the Low Level targets of reactors in both years' audits, you can notice that they even differ for the same reactor.

Anyway, the determination of High Level and Low Level targets is something which is connected with good audit practice. My audit experience told me that not every Team Leader uses these Target CLs. Usually Team Leaders feel that this is burden to them and they just try to have in their mind what the audit is about. Anyway, having a Target CL gives more formality to the audit, which is good from the point of legality of audits. But the Target CLs are not something that the Audit Team is submitting to the auditee (the company), so their legality can be questionable. Due to possible disputes after the audits, the Team Leader must prove that these Target CLs are produced before the audit. My recommendation is that when they

---

* Someone would say: "Integrating."
† These term "few years" may not be bigger than three years. Each auditor must strive to check the overall SMS in each company in two, and exceptionally three, consecutive years.
‡ Please note that I am not familiar with details about the operation of nuclear power plants. I am using it here only as example to show the connection and difference between High Level and Low Level Target CLs. Any implication that this is the way to do an audit on nuclear power plant is unrealistic!

are produced, the Team Leader must send them immediately to the Audit Team members.

The Regulator, by himself, should decide if the Audit Team will use them, but I am strongly recommending their use! In general, they will bring more benefit than harm.

6.2.2 Audit Check List

If you go on the internet and search for "Audit Check List," you will find a lot of hits. There are short and long tutorials about CLs and a lot of examples. But most of these examples are for ISO 9001, which is a Quality Management System (QMS). There are not so many CLs for SMS and for auditing Risky Industries, due to different regulations. Anyway, it is useful to take some model for CLs from the internet and adapt it to your regulations.

Let's remind you: The Audit Check Lists are CLs which are used during On-Site audits by auditors.

Whatever CLs the auditor decides to use, there must be at least three columns inside:

1. Clause or Paragraph from regulation – Here, the original number of the clause or paragraph, together with the original text, shall be mentioned.
2. Audit Question to be answered – Here, the question which the auditor must ask to get an answer on how the clause or paragraph is satisfied shall be written.
3. Evidence offered – Here, the response from the company (for the question above), together with the proof offered, must be written by the auditor.

I would like to present here one proposal (example) for an Audit Check List which I have used during a project with DGCA* of India (Figure 6.1). As part of my activities there, I was helping DGCA staff to deal with regulatory oversight audits. I will explain in detail why each column and row are used, and I hope it will help any Regulator to understand what each Audit CLs need to contain. You may use this one, or you may produce your own – it is your choice! But, have in mind that it is based on my experience, and I found this proposed CL extremely thorough and capable of eliminating all possible uncertainties during the audits.

The main rule regarding a good CL is: It is always better to have more things in the CLs than less. It will help the auditor and it will help companies to understand what is going on.

---

* DGCA stands for Directorate General of Civil Aviation. It is the name of the aviation Regulator in many states all around the world.

					Approval	Regular	Follow-Up	Exceptional		Special
Place:	**2**	Date:	**3**	Type of Audit	**5** ☐	☐	☐	☐		☐
Auditor:			**4**							
No.	Clause/ Paragraph		Text		What to ask	What is offered	CO NC CH			Remark/Comment
1	**6**		**7**		**8**	**9**	☐**10** ☐ ☐			**11**
2										
3										
4										
...										
...										
...										
Name of Auditor:		**12**			Date:	For the Auditee:		**13**		
Signature of Auditor:		**14**			**15**	Signature of Auditee:		**16**		

FIGURE 6.1
General Check List for audits.

As you can notice in Figure 6.1, there are fields marked with numbers. These numbers have different meanings and they help with the rules for filling the CL:

(a) Some of the numbers (6, 7, and 8) are about fields which need to be filled in the CLs which are *published* on the Regulator's website and they do not change with the audit, whatever the subject of audit is.

(b) Some of the numbers (1, 2, 3, 4, and 5) are explaining the fields which need to be filled *before* the particular audit.

(c) All other numbers are the fields which are filled *during* the audit. These numbers' fields are usually filled by handwriting during the audit and before the Closing Meeting, they are usually transferred to a computer in printable version. This printable version is printed as a document and it is submitted to the auditee for signing after the Closing Meeting.

Let's explain how and what to fill in all marked (by numbers) fields on Figure 6.1:

1. This is the title of the CL. Instead of the three dots, there should be the name of the company (unit, department, area, etc.) which is the subject of the audit. In addition, here could be mentioned the area of regulation (if there are different areas) which is the subject of this CL. For example: In aviation, auditing SMS of Air Navigation Service Provider, there are five areas of regulation: air traffic control (ATC), engineering services (CNS), meteorological services (MET), aeronautical information services (AIS), and search and rescue services (SAR). All of them are subject to different regulation and there is a need to have five Check Lists. In such a case, in each CL, this is the place where the area of regulation must be mentioned.

2. Here the place of audit shall be put. This is important for auditing a company where there are a few units (departments) in different places. The typical example for this could be an airport company handling a few airports, a power company having a few nuclear reactors in the same or a different place, a petroleum company with a few wells or rigs, etc.

3. This is usually the place where you put the date when the audit starts. Field 14 is the place where you put the date when the CL is signed by the auditor and by the company representative. Alternately, you may put in this field the period of conducting the audit (i.e., 12–16 November 2016).

4. This is the field for the name of the auditor and it is in accordance with the assumed reality that usually one auditor is associated with one CL. Sometimes, due to impartiality or to prevent complaints from the companies about one particular auditor, it is not wrong to assign two auditors for the same area. In such a case, both names should be here. Two auditors will attend the audit at the same time and they may produce two different findings for the same CLs. Later, they can discuss the findings during Audit Meetings and they will agree to produce common findings in one CL for this particular area.

5. This is the place where you tick a box for the type of audit. For Follow-Up, Exceptional, and Special Audits; maybe this format is not convenient, but if you like, you may produce your own check list for these types of audits.

6. Here, the number of paragraphs of the regulation (subject of audit!) must be stated. This field is strongly connected with field 7.

7. This is the place where the full text of the paragraph from field 6 is stated. It is important for the auditor to have this text, so during the audit, by reading the text to the auditee, he can clarify what and why it needs to be satisfied.

8. Here the things which need to be submitted by the auditee should be mentioned. It is some kind of guidance on what to ask. It can be some record, measurement result, list of material, recording, etc. What to ask depends on the text of the paragraph written in field 7.

Let's explain this a little bit in more detail:

For example: If the Regulator in the nuclear industry needs to approve a newly built power station, then there are requirements for Radiation Dose Limits (RDL*) inside and outside of the power plant. So, in this field should be written: Look for measurements inside and outside from independent source.

---

* Radiation Dose Limit (RDL) is connected with the other definition used where the explanation of how to measure it is presented. There is no need to present details here, but for more information (for a particular State), go to any website of any nuclear Regulator.

It means that the power plant's own measurements cannot be accepted. Of course, each nuclear power station must daily (continuously) check (monitor) the RDL inside and outside with its own measurement equipment, so regarding this requirement, own measurement results are accepted, but the auditor in this case must ask for proof of regular calibration of this measurement equipment in accordance with metrology standards. So, in such a case, there are two requirements: One for daily measurement results (ask for them and take a sample of few) and another for calibration of each piece of measurement system (check all of them and take a sample of the latest).

This field is very important from the point of the auditor and from the point of the auditee. Having in mind that this field is filled in advance and the CL is published on the website (submitted to the company in advance), the auditee will know what he can submit to prove compliance.

9. In this field you put what is offered from the auditee to prove that they are complying with the regulation from fields 6 and 7. It does not mean that the offered proof should comply with field 8 (What to ask). The Auditee may supply whatever they think can be used as proof, but the auditor decides if he will accept it or not. Anyway, the auditor is obliged to state the reason why it is not accepted in the Final Report.* Be careful: the Auditor must ask for a copy of what is offered to him! He will keep the copy as proof of compliance or non-compliance together with the CLs!

10. This is the field where the auditor just ticks one of the boxes. CO means *Compliant* and it is ticked when auditor is satisfied with the presented proof of compliance. NC means *non-compliant* and it is ticked when there is no proof of compliance. Usually this happens when it is obvious that the written procedure is not used or it is only partially used. For example: There are no records! CH means *Check* thoroughly later. Usually this is a situation when proof is offered, but there is a need for additional checking or additional consideration of the proof. For example, the record is offered as a proof, but the dates look strange and additional checks regarding possible fabrication of the record are needed. Maybe it looks strange, but it happens with all types of company's "personalities" sometimes.

11. This is a field which could be filled and maybe will not be filled. If the auditor notices something (good or bad), he can write it here. For example: Maybe the auditor asked for the copy and the auditee promised that he will provide the copy later, so the auditor will put a remark here that the copy needs to be submitted. Or the person who needs to explain compliance is not here at the moment, so the auditor

---

* The Final Report is a report regarding the audit results, and Check Lists are used to write Final Report.

will put a remark in this field as a reminder to come later again and check compliance.

12. The name of the auditor(s) should be put here. You are maybe confused because the name of the auditor is put also in field 4, but have in mind that the CLs have plenty of pages and it is good to have the name of the auditor near the place where he will put his signature. The name of auditor here, written by hand, confirms that the auditor has filled this page.

13. In this field, you put the name and title of the person who has accompanied the auditor during the audit. It should be person in charge of the area or department which is the subject of audit or a person who was assigned to the auditor by the company.

14. Here the auditor places his signature on the end of the audit in accordance with this CL. This field can be present on each page of the CL or only on the last page of the CL.

15. Here the auditor puts the date of finishing the audit activities connected with this CL.

16. Here the company representative (who accompanied the auditor during the audit) will place his signature.

6.3 What to Do with Filled Audit CLs?

When the On-Site audit for the particular day is finished, Audit CLs look very confusing. As mentioned before, the auditor is using the Audit CLs during the audit and they are filled usually by pen/pencil in handwriting. Even if a laptop or tablet is used, there is not enough time to put everything there. Usually the auditor fills the fields 9 and 11 with short notes or reminders expressed as acronyms, numbers, or notices. So, he will need another look later to put the CLs in order. This situation is well-known, so after each day of audit, there is dedicated time in the Audit Schedule when auditors can consolidate their CLs, samples of evidences, findings, and possible issues which are found during the audit. Sometimes even, there is need for some of these things to be discussed with the Team Leader and other Team Members.

Although the CLs are official audit documents, they are not submitted to the auditees. They are tools used to consider compliance or non-compliance and they are used to write the Final Report of the audit. So, they must be stored and kept together with all gathered samples, with all findings and with the Final Report in the archives of the Regulators. It is important to have them all the time, because looking at the CLs from the past audits will help the auditors to prepare better for the next audit. In addition, these old

CLs (from previous audits) were maybe filled by a different auditor and if this is the auditor's first time in contact with this company, it can help him to establish an opinion regarding the "personality" of the company (as stated in Table 5.1).

In the worst case, signed CLs can be used later in incident or accident investigations (if such a thing happens) or it is possible to use them in court as evidence in case of any kind of disputes of Regulator with the companies.

6.4 Compliance and Non-Compliance

As mentioned before, field 9 is reserved for the notes regarding the proof submitted from the company to the auditor that a particular regulatory requirement (mentioned in field 7) has been satisfied. Sometimes the particular proof submitted by the company is determined by the regulation, so the auditor know what needs to be submitted to him. Nevertheless, there is (maybe) a good reason for that and the regulation is a limiting factor for doing business for the companies (see Section 2.14, AMC).

Axiom 11: The good regulation should care only for effectiveness of the SMS! The Efficiency of the proposed solutions is concern only of the company, not of the Regulator!

Let's explain a little bit these two terms, effectiveness and efficiency.

In my humble opinion, these two terms are the best way to describe any company under consideration. It can be used by the company as a measure of success and it can be used by Regulators as way to establish an opinion about the "personality" of the company. Effectiveness and efficiency alone can be very subjective, because usually the company itself establishes criteria for effectiveness and efficiency. But associated with proper analysis of why this effectiveness (efficiency) was (not) achieved, it can provide an excellent picture about the "personality" of the company.

Effectiveness, in general, can be defined as a level of success of a particular process, operation, activity, task, etc. For such a purpose, it can be very simple to establish, or there is need for mathematics (statistics) to determine it.

Simple determination of effectiveness can be explained by the answer to the question: Did I reach my goal (Am I successful in what I am doing)? In such cases, there are only two possible answers: YES (I reached my goal, I am successful) or NO (I did not reach my goal, I am not successful).

Not-so-simple determination of effectiveness needs statistics as a tool, and in such cases, the effectiveness is expressed as a percentage. For example, the prediction of the CEO of the car manufacturing company was that in the next year, his company will produce 25,000 cars, but at the end of the year, the results said that the company has produced only 23,278 cars. So, the effectiveness of this company for the past year was only 93.11%.*

From the point of view of any Regulator regarding the audits, I would introduce the term "regulatory effectiveness" which can be defined by the question: Does the company comply 100% with the regulation or not? The value of 100% is very important in this case, because to get approval (especially in Risky Industries!), the company must be 100% compliant. In other words: The Company must comply with all regulatory clauses and paragraphs applicable for that type of regulation and that Management System. Regulators must not accept anything below this value.

Axiom 12: The Approval Audit is used to check if 100% compliance with the regulation is achieved by the company's SMS!

The logical question here is: Can the company be 107% compliant? And logical answer is: YES! This value has nothing to do with approval, because Regulator does not calculate effectiveness. What the auditor is doing is just checking are the all boxes CO in field 10 (Check List from Table 5.1) ticked. If all boxes YES are ticked, the auditor knows that the regulatory effectiveness for the company is 100%. But providing more than the regulation is looking for will show the attitude of the company to their "doing business" activities.

It is not strange at all! Sometimes a company may introduce its own requirements (specifications, tolerances, etc.) which are stronger than the regulatory requirements. From the Regulator's point of view, it means that the company have implemented something that is not requested, but there is no doubt that it improves safety. The reason for having such a situation is that very often regulation is a compromise between different levels of companies. Usually, States have brought in a regulation which will provide chances to do business for many companies and in such a way, the regulation is not so strong. But good companies may introduce their own rules (regulations, specifications, tolerances, etc.) which are stronger than those posed by the Regulator.

Maybe this looks to you like science fiction ("why implement something which is not required?"), but I had such a case in the 1997 in Frankfurt. I had a chance to join a study trip to DFS (Deutsche Flugsicherung), and in the presentations which were made to us, I noticed that their requirements for ILS (Instrumental Landing System) were stronger than those of ICAO, and I asked why they did that. The presenter told me that if they fail to satisfy these stronger requirements, they will be still good for ICAO requirements. With the simple words: The stronger requirements are pushing them forward to be better! And even if they do not become better, they are still good!

This is characteristic of a company from class 1 and 2 from Table 5.1. It will not affect the approval process, but it will very much contribute to building

---

* For the calculation, the following formula was used: 23278 / 25000 * 100 (%).

trust between the Regulator and the company. As I mentioned in Table 5.1, it is a pleasure to work with Generative and Proactive companies.

Efficiency is an economic category, and it can be roughly defined as the ratio of resources and production in the company. Let's say the company which is spending 200 KWh of electricity monthly for the production of one car* is more efficient then the company which is spending 210 KWh of electricity for one car. Maybe the difference in used electricity is not significant, but in summary, for all cars produced in one year (a few thousand in one year), it can become significant.

Anyway, the Regulator has nothing to do with efficiency, so the auditors even do not request data for it. Maybe the auditor could check the economic performance of the company from public data on the Stock Exchange as part of the data for establishing the "personality" of the company. But be careful: This applies only for an Approval Audit of merged or split companies. For an Approval Audit of a new company, these data are not available.

Efficiency is not checked during the audit, except if it is a regulatory requirement. But an experienced auditor can notice the efficiency of the company from documentation submitted during a Documentation Audit. Style and resources used during the implementation of proposed measures to comply with the regulation often give valuable information regarding the efficiency of the company which can help during the audits. Poor efficiency does not always mean that the company is bad. The company could just push itself forward (to be better!) and put more resources into their operations.

Effectiveness is part of compliance; efficiency could be, but let's speak here about non-compliance.

Non-compliance can be caused by systematic error or by random error.

Systematic error is error which, once introduced in the system, will always be present in the system. A simple example of systematic error is a watch which is adjusted (by mistake) to be five minutes forward. This error in the watch will be present all the time and the watch will always show the time as five minutes forward. Some people do that intentionally, especially in the morning, and they adjust their clocks to be five (or more!) minutes forward. This adjustment will provide to them a back-up time of five (or more!) minutes: If they are late by their clocks, they will be still on time by company's clocks!

In the scope of an Auditing Management System, systematic error can be:

(a) Missing procedure. (If there is a process and there is no procedure on how to conduct the process, then there is deficiency of SMS)

(b) Missing document. (If there is no document determining duties and responsibilities of safety personnel, then there is no legal obligation for each of them)

---

* Someone maybe will find the values used in this example as unrealistic, but this is not the point: I am just presenting simple example to explain efficiency.

(c) Missing rule. (If there is no rule regarding some situation, then everybody can do whatever he likes)

(d) Missing record. (If the record is missing, then legally: The activity was not executed at all.)

Each systematic error is non-compliance. It means something is missing in the Management System, so it is not holistic (100% compliance cannot be achieved!).

Random errors are errors which can happen, but there is no general rule how they happen. A random error can be:

(a) Some faults are not registered in the register, but there are notes in daily activities. (It was a simple fault and it was solved very fast.)

(b) Some reports are missing in documentation. (Some employees just forget to provide reports for different reasons.)

(c) Some corrective actions are not documented in detail. (The documentation is there in general, but the details are missing.)

(d) Some training certificates are missing. (However, the training was conducted!).

Random errors, if they do not happen very often, are not a reason for non-compliance. But if they happen often, they can become systematic error. For example, for the random errors mentioned above, I can say that if there are more than 20% of the documents missing or incorrect (unregistered faults, missing reports, lack of detail for corrective actions, missing training certificates, etc.), this is becoming systematic error. Especially if all of these missing documents are coming from one or two persons or happening in one or two departments, something is going wrong there...

Systematic errors can be easily classified as non-compliance, but the auditor must be careful with classification of random errors. Random errors could very easily happen due to Human Factors and this can happen very often due to a bad working environment in the company. This could be a real problem!

6.5 Audit Software

We are living in an Information Technology (IT) era and the effect of all these IT tools expressed by devices (laptops, tablets, mobiles, etc.) and the software applications which are inside them, cannot be neglected in our lives. The different software applications which are used on our gadgets really influence our everyday life, for good or for bad. I am not so much dedicated to all these

IT gadgets, and my sons and my wife usually mark me as "oldfashioned," but I cannot neglect the benefits which they provide.

The main point for this section is that, while IT gadgets are very much spread into our homes and our offices, I do not like to speak about their involvement in audits. Today there are plenty of different audit applications (software!) which are produced to help audits, but I do not like to speak about them in this book. This is not because I am "old-fashioned," but because there are a few other good reasons for such a thing.

The first reason is that while there are plenty of types of audit software applications (commercial and free for download on the internet), mentioning some of them in the book is simply not possible and not appropriate. It is not possible because there are too many of them, and it is not appropriate because it is not the intention of this book to do marketing for any software company.

The second reason is that these applications (whatever you are thinking about them) are only tools which can only help the audit process. They cannot replace the auditor, and everything that is mentioned in this book cannot be undermined by any type of software application which can be used during the audit.

And the third reason (and the most important!) is that whatever is written in this book has to do with all aspects of audit process and any of these aspects cannot be affected by any type of software application which can be used during the audits. Everything that is mentioned in this book is independent from any software application which the Regulator would decide to use during its oversight activities.

But I do not like to think that I am reluctant about using any type of audit software! On the contrary: I recommend it! But I must mention again: Each Regulator must not forget that this is only a tool which can be used during audits; it cannot replace a good auditor or an auditor's training.

The reason that I put this paragraph in this place is that the audit software can be (very successfully!) used for producing and maintaining the Check Lists. Whatever is said in this chapter and this book could be (and should be!) implemented in any type of audit software, so feel free to decide how you will use it.

But there is something else about the software applications. In general, each audit consists of three parts: Data collection, data analysis, and presentation of the results, and the software can be very useful during data analysis. The ordinary audit software deals mostly with structured data collection, CLs filling, and easy presentation of the results, but for data analysis, it cannot provide a benefit.

So, there is additional software dealing with something else which is different from audit software. I am speaking about Safety Management where the Risk Assessment must be done using some of the methods for Risk calculation. Almost all of the methods for Risk Assessment can be done by some software application (different from the audit software), and these applications are very useful. They could be applications dealing with some of the methods or methodologies for Risk Assessment, or simply software for statistical calculations.* These types of software are inevitable because without them, the Risk Assessment and data analysis will need too much time and will produce questionable quality of the results.

---

* Statistics can provide very good results in auditing Safety Objectives, preventive and corrective maintenance/actions, processing hazards and risks, safety events and safety performance.

7

Management of Findings

7.1 Introduction

Dealing with the findings of audits at this stage of the book perhaps to someone will look too early, but I found it very important. Audits are about the findings and the results of audits are presented as findings. Explaining the findings here will help auditors to understand later why the particular action or activity during the audit is done. I do believe that things regarding audits that are mentioned later will be easily understandable if you read this chapter now.

7.2 Findings

During the audit, the auditors are looking for an assurance that the regulation is satisfied, and this assurance is the most important result. The overall activities during audits can be defined as checking the compliance of the Management System with the regulatory requirements. If there is a requirement for a procedure for everyday monitoring of equipment, the auditor shall look for proof that the equipment is monitored every day. He will ask for proof and probably the lists of monitored notes will be offered for him for each day. If he notices that there are days on which monitoring was not done, then it is not good, but it does not mean that monitoring did not happen. In such cases, the auditor must proceed carefully.

Axiom 13: Whatever you think about the audit process, an audit can be defined as a quest for compliance! Whatever you think about the audit process, it is not a quest for non-compliance!

The auditors must be assured that the good things are present. It is possible not to have good things and not to have bad things. But the presence of bad things does not mean that good things are missing. To clarify this, I can use a simple example:

> You do not evaluate a good football player by the chances to score goals! You evaluate him by the numbers of goals scored and by the number of assists when others are scoring goals. Of course, you do not like misses, but you value good things attached to this player which are contributing to the overall team performance.

There is something in human nature which is more dedicated to bad things than to good things: Humans cannot provide good value for good things, because they do not hurt them, but humans remember the bad things (they hurt!) and there are novels written about them!

There is also another thing which is important for auditing: The lack of bad things does not mean that there is the presence of good things. A regulation is made with the intention to reinforce good things and to eliminate bad things. In every audit, you can find some errors or mistakes, but it does not mean that there is a bad Management System. So, the auditor is checking at first the presence of procedures (actions, measures, operations, tasks, etc.) which are building the Management System. After that, he is checking are these procedures (actions, measures, operations, tasks, etc.) implemented, followed, maintained, and improved over time.

In a more detailed description, I can say that there are two phases of each type of audit:

(1) Documentation phase (represented by Documentation Audit), when the auditor is investigating if the documented procedures (actions, measures, operations, tasks, etc.) provide a way to satisfy the regulation.

(2) Checking phase (represented by the On-Site Audit), when the auditor is going on-site in the company and he is checking the implementation and following of the procedures (actions, measures, operations, tasks, etc.) in daily operations.

7.3 Objective Evidence

In both phases from the section above, the very important thing is to provide objective evidence that compliance with the regulation is achieved. The term "objective evidence" can be defined as data (in the form of records, written statements or commands, measurements, reports, etc.) presented to the auditor which will prove that the company is doing what is required by the regulation. Some of these documents are important for the functioning of the company, so if auditor cannot get them in such a case, he may ask for a copy of them. The main point for gathering objective evidence is the fact that the auditor is obliged to prove (to himself, to the company, to the Regulator, and to the public!) the integrity of findings which are reported.

If you read carefully the definition of objective evidence, you will notice that the words opinion, impression, view, belief, or judgment are missing. The auditor may not state, or comment on, something which is based on his opinion. He may have an opinion about the things (tasks, procedures, activities, operations, etc.), but he must keep it for himself.

Axiom 14: The auditor must not under any circumstances express or take into consideration his own opinion about the things (tasks, procedures, activities, operations, etc.) in the company which is the subject of the audit! All his statements or comments must be based on objective evidence!

By stating his opinion, the auditor could bring the impartiality of his work into question, and it can be used later against him if there is a dispute regarding the audit findings. Also, it can be very, very offensive for the company.

Usually a company can offer evidence that the procedure has been implemented and followed, but this is something which must be written in the SMS manual. So, the auditor may ask for these records which are part of the manual. If the auditor is not satisfied with the evidence, he must ask for more or remind the Counterpart (company representative!) what is written in the company's manual. Using the CLs from Figure 6.1, field 8 should contain the list of acceptable evidence. To build this list, the Regulator may use the usual kinds of evidence which are offered to him or which he finds acceptable.

It is wrong if the auditor is limited to the list from field 8 of the proposed CL. The issue with the objective evidence is not only its presence, but also the quality of the offered evidence. Having in mind Axiom 5, during checking of the document, the auditor must understand how the company's system is functioning and based on this understanding, he can ask questions and he can evaluate the quality of the offered objective evidence.

So, we can say that there are a few issues with the objective evidence:

(a) Is the evidence offered?
(b) Is the offered evidence in accordance with the one written in the manual?
(c) Is the offered evidence objective? (Or can the auditor get a clear understanding of what is going on with this evidence?)
(d) Is the evidence in accordance with other evidence offered in different areas during the audit? (More about this in Section 9.2.12!)

The last one is worth considering.

During the Approval Audit, the overall SMS is checked, and the overall SMS is complex, as doing business is a complex thing. So, there are a lot of interactions and correlations between the processes in the company which have to be reflected in the SMS. It means that evidence for the implemented procedures or procedures is also interactive and correlated. In addition, sometimes evidence offered can be in conflict with each other. There are plenty such examples, and just for clarification, I will mention one which was already mentioned in Section 3.8 in this book: The *Deepwater Horizon* disaster in 2010. There, this disaster was mentioned as example of poor regulatory Oversight (this poor Oversight was even mentioned by the US investigator in the Final Report), but I will use it to make a point in the areas of other objective evidence.

There was a process to close the spill in the British Petroleum (BP) rig (named *Deepwater Horizon*) and there were the measurements which they needed to do. They did the measurements, but the results of the measurements were inconclusive. Actually, the results did not show that the rig was OK, but the BP experts decided to continue. Obviously, they did not follow their procedures and it resulted in a huge accident.

And this is the area where a good auditor will compare the offered documents regarding the functioning of the SMS: Measurement results and Follow-Up actions. To check the objective evidence that the process of any measurement is followed, the auditor will check for measurement results which are not in line with the requested results. The company must investigate why the results do not comply with the standard for such a type of measurement. If there is proof that the results were not good, the auditor must check what was the FollowUp of the bad result. If the SMS is OK, there must be a method to investigate the bad result and some action (Preventive or Corrective) must be implemented.

This happened to me when I provided On-the-Job Auditor training for five trainees of DGCA in India. I was on an On-Site Audit and I was checking the preventive maintenance data for VOR* (VHF Omnidirectional Range) installed on the aerodrome in Mumbai. I noticed that three weeks earlier, the VOR had some problems with the signal (it was not within ICAO tolerances) and I asked them for proof that this bad signal was investigated (what they did to understand why the VOR had a problem with radiation). They told me that they did nothing. It was a Level 2 finding.

What is the point here: There was evidence that something is wrong, but there was no evidence that something was done at least to investigate what is wrong! There was no activity which was followed in these cases, which means that the evidence did not comply with each other.

There is another case in these situations where you are checking the "Chronology of Events." It is case of conflicting evidence.

A simple example for this case can be one when the auditor is checking the records of measurements for radioactivity in a nuclear power plant and all of

---

* VOR is navigation equipment which provides a signal to aircraft in flight regarding the azimuth of the aircraft from the VOR position.

them are done by one guy. But checking the dates of the presence in the power plant of this guy, the auditor noticed that he was on leave for two weeks. So, the guy was not present and measurements were only done by him...

This is obviously an event which could not have happened in these two weeks. So, the auditor should check the reason for this. Possible reasons are:

(a) The measurements were falsified. (Worst case scenario!)

(b) The measurements were done by someone else who was not competent for this. (Also a worst case scenario!)

(c) The guy canceled his leave and he did the measurements, but HR did not update the documents. (Less bad scenario.)

Conflicting evidence is not always a cause of problems. Statistics say that usually a little deeper investigation can clarify the reason for the conflicting evidence and it can show that this conflict is often not a conflict at all when the auditor looks at the culture and complexity of processes in the company.

Anyway, these are things regarding the integrity of the objective evidence which must be clarified by the auditor during the audit.

7.4 Type of Findings

During the audit, the auditor could discover a lot of findings, but not all of them will bear the same importance. So, there is a need to establish criteria on what is important (What can endanger safety?) and what is not important (It may not necessarily endanger safety!). Different Risky Industries have different criteria (mostly three or four of them), but there is some average level of findings' importance which I will present in this book.

In Table 7.1 are presented three levels of audit findings which are extracted and combined from different Risky Industries. The purpose of presenting these findings in the book is to show the auditor how to use them while writing the Audit Final Report. Their level of importance (significance) will trigger action by the Regulator and by the company later. Whatever the level of the finding is, it must be accompanied by objective evidence.

If you go on the internet, you can find a lot of such examples regarding different levels of findings, as presented in Table 7.1. There are examples with five levels (Critical, Major, Moderate, Minor, and Observation) and there is nothing wrong with those. In my humble opinion, the three levels as mentioned in Table 7.1 provide more flexibility. But whatever the level of finding is, it must be defined in the regulation. So, the particular international organization and/or the Regulator of the State may produce different levels of findings, but they must be part of the regulation in that area.

TABLE 7.1

Categorization of Audit Findings*

Level	Name	Explanation	Action
1	Major	This is non-compliance with one or few regulatory requirements. It can be: missing procedure, incomplete procedure, missing record, missing training, missing document, missing implementation of something, stopped following the procedure, non-implementation of preventive or corrective action, non-implementation of regulatory requirement, etc. This type of non-compliance is serious because it may endanger safety standards and safety performance. Going further (depending on the type of audit and nature of non-compliance!), the Regulator may issue a fine or even suspend the particular operation which is the subject of this non-compliance!	The company must comply with the regulation. The Audit Team will give the company a particular period of time to solve non-compliance. The time period given will depend on the nature of the non-compliance. If the procedure is missing, they can give time having in mind time needed to provide the procedure and to train the staff. Whatever the non-compliance is, the company may not be approved until they do rectify this finding. They must fix the problem as soon as possible! The Regulator will provide a Follow-Up Audit later for this finding only when the company reports that the non-compliance has been solved!
2	Minor	This is a lower non-compliance compared to Level 1 and usually is connected with an activity which do not endanger safety performance immediately, but it lowers the safety standards. It could be a missing record for an activity which has already been done or maintenance already executed, missing training, or record for training. It means that something is missing and it will not produce a safety risk at the moment, but if not treated, later it can transform itself into Level 1 non-compliance, or it will result in an incident or accident.	The Audit Team will give the company a particular period of time to solve this non-compliance. The time period given will depend on the nature of the non-compliance, but not more than three months. Whatever the non-compliance is, the company must fix it! The Regulator will provide a Follow-Up Audit later for this finding only when the company reports that the non-compliance has been solved!

(Continued)

---

* This table is presenting regulatory levels of audit findings for the SMS in the companies which are used in aviation. Anyway, it is my representation which is in line with the official ones. In all other industries, there should be similar categorizations which need to be used during audits.

TABLE 7.1 (CONTINUED)

Categorization of Audit Findings

Level	Name	Explanation	Action
3	Observation	This is not non-compliance. This is just a notice from the auditor. Anyhow, this is very disputable level of finding. There is a fine line between the personal opinion of the auditor and the finding which, if not treated, could produce a safety issue. The auditor must be careful with these observations!	No action required in such a case. Companies Class 1 and 2 (maybe 3) will voluntarily consider this finding. Other companies can do it, but the Regulator may not push it. If the Regulator thinks it is important, then it can change the regulation and he must provide training for this change.

For each level of finding, company may provide their opinion and it very often happens that they simply dispute not only the finding, but also the level of finding. The auditor may be careful in the way in which he will present the findings and the objective evidence. It is good if he can recall some other (same and/or similar) situation from the past which happened here or in some other place around the world. If a recommendation on how to deal with such a situation exists, it may be used and the company must have the chance to become familiar with it.

All the findings are important information regarding the safety performance of the company, and if there is such a concern regarding Safety, the auditor may use his power to compel the company to solve the problem. This right should be used only for Level 1 and Level 2 findings.

The general advice to the auditors in a situation when the dispute about the nature or level of findings exists, is not to rush with solutions. Do not forget that immediately after the audit, the Team Leader presents just the preliminary (interim) report, so it is wise to accept all comments, notices, and complaints from the company during the Closing Meeting. A few days later, the Audit Team can meet again to reconsider the overall situation. If necessary, the Team Leader may try to consult with other auditors in the Regulator (who were not present on the audit!) or with the Specialists.

But there are situations which need an immediate and strong response!

If the auditor finds a Level 1 finding which present immediate danger for Humans, premises, and the environment, he may not wait. Immediately (in consultation with the Team Leader and other auditors), he must raise a Level 1 finding and request (strongly!) the subject of this finding to be ceased:

(a) Procedure shall stop being used and an intermediate procedure shall be produced.

(b) The operation, activity, or process shall immediately be stopped.

(c) In the case of significant non-compliance (dangerous behavior!), the company may even be closed.

Such a thing is a Level 1 finding, which could be:

(a) Radioactive material is stored outside without any protection.

(b) Dangerous chemicals are dropped in the river or soil.

(c) The aircraft is not fit for flight.

(d) Pilot or ATCO is drunk or drugged.

These are things which are an immediate Risk for Humans, assets, and the environment, so there is no mercy there!

7.5 How to Present the Findings?

In general, the first "glimpse" of findings is presented at the end of the audit, through the Interim Report at the Closing Meeting.

Good practice says that the Team Leader could start with the good things which were found during the audit. Do not forget Axiom 13 (Audit is about good things; it is about compliance!). Auditors, through overall audit activity, are looking for implementation of systematic solutions which will satisfy the safety regulation in the particular Risky Industry. So, these good things should be presented at the Closing Meeting first. And this is the only place where every good opinion, good notice, or good impression gathered by auditors can be expressed. For everything else, the auditors need objective evidence!

After that, the Team Leader may explain what the findings are, how they are connected by regulation, and how the level of non-compliance was determined. In addition, he will ask how many days (weeks, months, etc.) the company will need to rectify all the noncompliances. It should be a firm request, but it shall be presented politely. It is good practice to accept every *reasonable* proposal from the company regarding the time needed to fix noncompliance. The emphasis is on the italicized word "reasonable"! The Team Leader must have in mind the Level of findings: The Level 1 findings could require immediate action!

Depending on the company's reaction, the Team Leader will decide how to proceed, but there is no need for rushing or fast responses. Whatever is said in this meeting could later be written in the Final Report. In any step and at any time of this process (of writing Final Report), the company may provide its opinion.

7.6 What Is Next?

After finishing the audit with the Closing Meeting, the auditors leave the company. They may take a few days off* and they should meet again in Regulator's premises. They will have a Team meeting when they go through the preliminary Final Report again and, based on the discussions, the Team Leader will produce the Final Report.

Together with the Final Report of the audit, there shall be a letter from the Regulator with the period of time given to the company when they need to propose the measures or actions to handle the findings. This period of time should be reasonable and should be based on the nature and complexity of findings.

The company must respond to the requirements in this accompanying letter as stated by the Regulator, and they need to submit proposals with proposed measures (Preventive or Corrective Actions) for solving the non-compliances expressed as findings. Depending on the nature and complexity of the findings, the company may respond by a simple letter or by building a Safety Case. If they need to build a Safety Case, they must send a letter to the Regulator immediately to explain the reason for that. The Regulator must analyze the proposed measures and/or actions by the company stated in their response and they may or may not approve it. But this process of "Approval/Non-Approval" could be very tricky.

Regulatory Auditors must understand that companies are doing business, so they cannot run away from the economic impact of the measures/actions which they propose. Depending on their "personality," a different approach can be experienced by the Regulators. The measures/actions which the company will propose will be in the form of a Safety Case and it shall be accompanied with the particular analysis of the risk. Which method for this analysis will be used depends on the regulation or, if not precisely required by the regulation, the company may decide itself how it will analyze the risk.

Anyway, the Regulator must have the capability to check the Risk Assessment for these measures/actions and if it is not clear, they may ask for additional clarifications. To be honest, it is not always possible to get a clear picture from reading the Safety Case and Risk Assessment. If the Regulator is using their own method (or methodology) for Risk Assessment (different from the company), auditors may use data from company's Safety Case and they will check the outcomes with their method. If there is good compliance, then they may approve the measures/actions. But if there is no good compliance, the Regulator must understand why there is non-compliance.

---

* It does not mean that they should go on holiday! They can stay working at the Regulator's premises, but they can dedicate themselves to other activities. It is good for your brain to take a rest from the passed audit for a while.

Maybe the data were not applicable to the method used or some of the data are missing, etc.

The company has the primary responsibility to provide Functional Safety of their products and the services offered, so this must be taken into account. The Regulator may not be happy with the chosen solution, but he must provide *objective evidence* why this solution will not work. Again, auditors may not make decisions based on their opinion!

I remember that on one of the safety courses which I attended a long time ago, a very interesting example was given to us:

The company submitted the Safety Case for the change in the operation which they needed to improve. This was not a standard operation and it was not regulated by the regulation. So, the Regulator needed to validate the company operation through Approval. Although the company provided a considerable amount of data, the Regulator could not determine if the operation was safe. The problem was that they could not even determine that the operation is not safe. Whatever analyses they did, the results were inconclusive… And this was not enough to be presented as objective evidence!

So, the Regulator sent a Letter of Approval to the company where it was written:

> We were not convinced by your Safety Case that the change of operation which you propose is safe. We also could not be convinced that the change of your operation is not safe. So, having in mind that the primary responsibility for safety of your operation stays with you, you may implement the change on your own responsibility!

This is a good example of how the Regulator should behave in cases where objective evidence is missing.

Whatever the decision of the Regulator is, implementation of Corrective or Preventive Actions must be checked by the Regulator. It is done through a Follow-Up Audit. As was explained earlier, after the Approval was sent to the company regarding their Safety Case, the company was given time to implement the measure. In the same letter with the Approval, the company was told on which date the auditor would come for the Follow-Up Audit and that he would check the implementation.

The Follow-Up Audit can be done only by one auditor and there is no need for to prepare some program or use some Check Lists. If the proposed action is complex, then the auditor could prepare a Check List on what to do, but this is not submitted to the company in advance. This is only for the auditor's purposes. Depending on the complexity also, the Follow-Up Audit can last a few hours or a few days. If it is complex, maybe one auditor is not enough, so there is need for a few auditors or even for specialists. Anyway, the Team Leader will decide how to proceed.

8

Preparation for Audit

8.1 Introduction

There is a schedule of activities during the audit and it will be explained here. First Party Audits (Internal Audits) and Second Party Audits are not part of regulatory audits, so they will not be subject to consideration in this chapter. From the Third Party Audit, only Approval and Regular Audits will be explained here. The other three (Follow-Up, Exceptional, and Special Audit) are just simplifications of the Approval and Regular Audits, so they can be done easily on case-by-case basis by anyone. I will point mostly to the requirements regarding the Approval Audit, and whatever is said here, but simplified, can be used for the Regular Audit.

If the Regulator would like to provide a good audit, he must prepare himself for the audit. The preparation means that there is a need for establishing the Audit Team with a Team Leader. The Team Leader will produce the audit plan which will be expressed through the Audit Schedule. The Audit Schedule (activities and duration of audit!) and number of the Team Members (auditors) should be produced taking care for the scope (type!) of audit. These three things are correlated and they are presented in Figure 8.1.

The scope of the audit can be defined as a measure for quantification and quality of the details which will be considered during the audit. Speaking about the details here means how deep the audit will be, if it will cover everything in each department,* or if it will focus just on particular areas (see Section 8.5, Sampling). The scope depends on the type of audit. The Approval Audit has the biggest scope, and it is valid especially valid for the Documentation Audit. During this audit, every document submitted by the company is checked and there are no exceptions. Later, during the On-Site Audit, not everything is checked. Follow-Up, Exceptional, and Special Audits have a scope which is limited to one or few activities. The factors which can affect the scope of audit are:

---

* By the "Rule of Thumb," there is no regulatory audit which covers "everything in each department;" this is done usually in the case of accident or incident investigations.

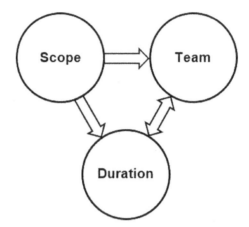

FIGURE 8.1
Correlations between scope of audit, number of auditors (Team), and duration of audit.

(a) Type of audit
(b) Size of company
(c) Regulation requirements
(d) Previous audits
(e) History of Safety Events with the company
(f) Reliable outside information

The Team Leader will determine the scope of audit, and he will proceed by determining how big the team will be and how many days the audit will last (duration!) A bigger scope will result in more team members (auditors) and more days for audit. Increasing the number of auditors can decrease the days needed, but this cannot be always done. The number of auditors inside the Regulator is limited and it is based on the number of employees. The Team Leader can decide how many of them will be engaged for a particular audit. The number of team members will determine the number of days. The main rule is: Make the OnSite Audit as short as possible! The Documentation Audit may last as long as necessary!

As can be noticed, there can be adjustment by the Team Leader regarding the resources for the Documentation Audit and for the On-Site Audit, but it is a "Golden Rule" that the auditors engaged in the Documentation Audit shall continue with On-Site Audit. The reason is very simple: Whatever the auditors "encounter" in Documentation Audit, they will know how and where to check during the On-Site Audit!

It is important to understand that fewer days auditing will provide a Win-Win situation: The Regulator will benefit from fewer days with one company (the auditors will have more time for other audits) and the company will have more time to dedicate to its operations. Unfortunately, the compromise must be made by the Regulator. The limiting factor there (as I mentioned above) is the number of auditors employed in the State Regulator. There is no recommendation on how to do it and each Regulator must decide how many auditors need to be employed and how many of them will be used during the audits.

The Regulator must take into consideration the following aspects regarding the audits: Type and numbers of companies (subject of regulation), history of previous events, type of audits, objective of audits, etc. Based on these aspects, the Regulator will prepare a list of resources needed for audits and a yearly plan for audits. Also, the audit is just part of the regulatory oversight activities of Regulator, so it is good practice if the number of determined auditors is increased by 30% with specialists, drivers, accounting, legal, and other administrative staff.

8.2 Schedule of Audits

Regulators oversee a lot of companies, so they must prepare themselves for each audit considering available resources (how many auditors?) and timing (when and how long the audits will last?). Based on their activities and number of companies overseen, the Regulator needs to employ and train particular number of auditors. Assuming that such a number of auditors is available, there is a need to establish an annual Schedule of Audits for the next year. It is good if this schedule is sent (in advance!) to all companies which are subject to audits and, in addition, it is good to publish it on the website of the Regulator. It should be done in September at the latest (for the next year starting in January) and there are two reasons for that:

> The first reason is that for all Approval Audits, first there is a Documentation Audit which precedes the On-Site Audit by two months usually. The second reason is that it is good to inform companies in advance, because in that time they are also planning their activities for the next year. Do not forget Axiom 8: Regulators and companies are partners in achieving safety!

This annual Schedule of Audits covers the annual audit activities of Regulator, but it is a live document. It means that must be stated that the Schedule of Audits is interim document and it is flexible (subject to regulatory requirements!) Do not forget that some events cannot be planned, especially bad things: If there is an accident, usually all planned audits are postponed and everything is dedicated to the investigation and elimination/mitigation of the consequences of this accident.

8.3 Establishing the Audit Team

Before the audits, there is a need to choose a Team Leader* who will establish the Audit Team.

The Team Leader is a *trained* person who has proven himself to have the necessary knowledge, experience, skills, and attitude in the particular area. The emphasis is put on the italicized word: *trained*. Whatever is his expertise in particular Risky Industry, the Team Leader must be trained in regulatory audits! Actually, this is requirement for all Team Members (auditors) from the Audit Team: Each of them must be trained for conducting audits!

Axiom 15: Expertise in a particular area is not the only requirement for the Team Members of Audit Teams. Equal importance is given to the requirement to provide proper auditor training for each Team Member!

Usually the Team Leader is the person who will determine the resources and duration of the audit. Depending on the size of the company which is the subject of the audit and the volume of applied regulation, he will choose a particular number of Team Members. The Team Leader may divide out the received document(s) to the Team Members in accordance with their specialties and areas of expertise. The Team Leader also does the audit (he is a member of the Audit Team!), but the volume of his audit activities is not so big as the ones of other members of the Team (auditors). The reason for that is that he is actually the connecting link between the Regulator and the company. If there are any requirements from the auditors, they will be delivered to the Team Leader and he will submit these requirements to the person responsible for audits in the company. The same thing applies also for any disputes: He is in charge of resolving them with the person responsible for audits in the company.

Establishing a Team is necessary because the audit is a complex job. When the Regulator oversees the Safety Management System (SMS) of a company in Risky Industries, there are many areas of concern which needs to be checked. It is unrealistic to expect that one person can do it. OK, one person maybe can do it, but it will last few months, and I really doubt that it will bring any useful information. In addition, each company (by itself) is complex entity and the auditors must take care for that. So, each of these entities needs different expertise and there is a need for Team Members with particular expertise in requested areas.

Team Members (auditors) are persons who actively take part in the audit, but if there is a need for additional expertise, the Team Leader may engage

---

* In some places, you may find "Lead Auditor" instead "Team Leader." I prefer Team Leader, because Lead Auditor is a title which you get when you pass the exam for Lead Auditor. The Audit Team may have few Lead Auditors, but only one Team Leader!

Specialists. The Specialists are experts in particular areas who are not trained as auditors and they are not present onsite (in the company) during the audits. They can be employed by the Regulator or they can be engaged on contract at a particular time. If there is a particular need for their presence during audits, they can be invited, but this is uncommon for Approval Audits. For Exceptional, Special, and Follow-Up Audits, they can be invited, if their presence is necessary. The Specialists can also be engaged in situations where the particular Safety Case is submitted to the Regulator for approval. If the Safety Case is regarding implementation of new technology, then the expertise of the Specialists is essential for understanding the safety impact of this new technology in the operations of the company.

Remember what is written in Section 2.1 (Introduction) in this book: The Auditor does not need to be a chicken to lay an egg, but he must have the capabilities and knowledge to determine which egg is good and which one is bad!

In cases where there is a need to know "how the egg should be laid," auditors will engage Specialist(s).

More details about the personality of Team Members (auditors) will be provided in Chapter 11 (Profile of Auditor).

8.4 Organizing the Approval Audit

As it has been mentioned in Section 4.3.2.1 (Categories of Third Party Audits), the Approval Audit is an audit which is conducted on the newly established company and it consists of a Documentation Audit and On-Site Audit. In addition, this is an audit which is executed when there are situations where two companies are to merge or one company is to split in few other companies. Remind yourself of Axiom 3: There are no two of the same Management Systems implemented in different companies!

In the case of merging two companies and/or splitting one company to few, there is considerable change of structure of the company. The change of structure will trigger considerable change in organization of the company. In total, a new Management System must be established and this new system will cover the new reality of the company. The new organization and structure of the company will need new Management Systems and it must be the subject of regulatory audit.

Whatever is the case from above, the new companies must submit the request for registration and approval from Regulator. In this request, they will submit a considerable amount of documentation* regarding the com-

---

* In aviation, this documentation is known as Company Exposition, and each aviation subject must submit it to the State Regulator.

pany which will prove that the company is fulfilling requirements for safe functioning in the particular Risky Industry. Part of this documentation will be the SMS manual* and this manual is usually subject of the audit: Auditors check if the regulation is satisfied and if everything in the manual has been implemented.

The company will send the documentation on to the Regulator's address and it will be given to the (nominated by the Regulator!) Team Leader. He will read the documentation and he will distribute documentation to other members of the Team in accordance with the areas of their expertise. First reading of the documentation by the Team Leader should give a clear picture about the size and expertise of the members of the Audit Team.

This is the time when the Documentation Audit starts!

In addition to the data submitted from the company, to prepare themselves for audit, the auditors may use other sources of data regarding the company. In the case of a Regular Audit, there must be data from previous audits. Maybe there are not so many audits before, but at least the Approval Audit was done one year ago (depending on the industry!) This process of checking into the company is sometimes called Intelligence. As spies gather information from different sources, the auditor may gather data from previous events, submissions of Safety Cases, or previous audits. In general, there is a need to build a picture of what can be expected from the company during the audit.

There is nothing strange in this process! Just as sports managers are watching (live or on video) matches of their opponents to prepare for the following matches, the same applies for auditors: They are using Intelligence to prepare themselves for the next audits. But also, as managers can adjust the team play during the match, so the auditors must adjust their activities during the audits.

The picture built for the company before the audit is just an interim one and it could be changed before or during the audit. The auditor may not use this picture as prejudice about the company! This picture will provide to the auditors only the expectations on how the audit could be conducted.

8.4.1 Documentation Audit

The company's SMS manual is used as a help to the Team Members to get familiar with the company and with the information on how the SMS is implemented in the company. The first thing which needs to be determined by the Team is: Does the Manual comply with all regulatory requirements for SMS? The particular amount of time needed to do it depends on the volume

---

* The Manual is a requirement for each Management System, but the structure of how it is produced is not always the same. It could contain inside the procedures or it may just point to the particular procedure which exists as an individual document. In this book, I will stick to the first case where the Manual is the only document which completely explains the implemented SMS by the company.

of the company. The time period can vary from one week (for small companies) to two months (for big companies). It also depends on the number of auditors engaged in the Documentation Audit.

The Documentation Audit is a very important part of overall oversight process, because this is a place to check and to find any systematic error in the SMS, if it is present. During the OnSite Audit, mostly the individual and random errors can be discovered.

If there are things in the documentation which are not clear or there is any type of uncertainty regarding particular solutions (operations, procedures, etc.) implemented by the company, the Documentation Audit is the time when it needs to be clarified. Depending on the audit organization, all Team Members will submit their reviews and questions to the Team Leader. He will summarize them in one official letter to the company and the Audit Team will meet to discuss this official letter. When the official letter is agreed, the Team Leader will send it to the company. Usually this is the person nominated from the company as Point of Contact. If there is no such a nomination, the official letter could be sent to the General Manager or Safety Manager in the company. The Team Leader is at the same rank as the company's Safety Manager, so I can recommend that the official letter be sent to him.

Depending on the nature of requests for clarification, the company can be given one or two weeks to respond to the official letter. Anyway, this time is not critical for Regulator, but it is critical for the company: As soon as they respond, they will speed up the Approval Audit process and they will soon start with their operations.

The response may answer some of the questions and clarify some of the uncertainties, but if it shows some non-compliance, this is time to request changes, upgrades, updates, or whatever is necessary. Depending on the requests sent to the company, it must be given enough time to them to rectify the non-compliance.

The response from the company regarding non-compliance issues is actually the first insight into the company "personality:" The level of time spent to provide a response to the Regulator's requests and the content and clarifications offered in the response will give the auditors the first hint what type of company it is.

The response can be analyzed by the Team Leader, but it must be discussed in the internal Team meeting. When each Team Member is satisfied with the responses covering his area of responsibility, then the Documentation Audit is finished: The Documentation (the SMS Manual) will be approved and the date(s) for the On-Site Audit may be proposed!

Axiom 16: If the Audit Team is not satisfied with all answers during the Documentation Audit, the company's SMS manual must not be approved!

Axiom 17: If the SMS manual is not approved, there is no reason to proceed with the On-Site Audit!

But there is a question: What if the overall process of providing company's response is consuming too much time?

In general, in regard to the Approval Audit, it is not the concern of the Regulator!* Each company starts with the intention to offer some product or service on the market. This product and/or service will bring profit to the company, so going early to the market means the profit will be bigger. It means that the interest of the company is in starting its operations as soon as it is possible, so the company needs to struggle to obtain all licenses, certificates and approvals from the Regulators quickly and to register themselves as an entity as soon as possible. But, having this in mind and having in the mind Axiom 5 (The Auditor must understand how the company is trying to comply with the requirements!), it is the duty and unwritten obligation of Regulator to provide regulations and conditions for the companies to be able to access market as soon as possible!

Anyway, if the Regulator does not receive in timely manner responses for their questions regarding the Approval Audit, there is no need for concern. There are plenty of cases where the companies intentionally delay their responds and the reasons for that can be different:

(a) They are not ready with something else, so their focus is dedicated to something else.

(b) They have noticed from your questions that their SMS has deficiencies and they need time to solve these deficiencies.

(c) They simply do not understand the questions posed to them and they are looking for proper understanding. This "proper understanding" can be offered from someone else and they are engaging some other company or external consultant to help them.

Whichever of these reasons apply, the final product (company's SMS!) will be given a chance to be better. It means that there is no reason for the Regulator to jump to wrong conclusions and the Regulator should not push the company to react immediately.

8.4.2 Preparing On-Site Audit

Each piece of documentation submitted by the company, however how big or small it is, must satisfy all requirements for the Documentation Audit. The same type of documents must be produced by each company and they will differ significantly by the context and by size. I can say that the Documentation Audit is the same for each company, regardless of its size. But the On-Site Audit differs. It is strongly dependent on the size of the company.

---

* For the Regular Audit, the response from the company must be received in the time period determined by the Final Report!

Bigger companies have different business issues and challenges than small companies. With big companies, the overall structure and organization is more complex and it will be reflected in the implemented Management Systems. Also, the big companies will need more auditors in the Audit Team and small companies will need fewer auditors. The Team Leader must take care of it.

When the Documentation Audit is over, then the Team Leader will determine the length in days of On-Site Audit and he will produce the Audit Schedule. He will send a letter* to the company with the date of the On-Site Audit together with the Audit Schedule. In the letter, he must state how many auditors are coming and he should ask for a room with internet access, which can be used during audit only by auditors. This is the room where the Audit Team can meet and there they can consider and comment on gathered data during the audit, discuss findings, possible non-compliance, uncertainties, etc.

The company will respond to these letters accepting the date and the Audit Schedule or maybe they will oppose the date or the schedule.

There is nothing wrong if they oppose the date and/or the schedule. Do not forget that they have business to do! Regarding the date, maybe there are operations which will prevent having an audit on this date or maybe the responsible persons (Safety Manager or someone else) is sick or on leave on these dates. Regarding the objection to the Audit Schedule, the reasons may be similar or same.

When the company accepts the date of audit, they must inform their employees who will be affected by the audit and they must prepare themselves. There should not be surprises during the audit between the employees and the auditors.

The Team Leader should consider all these reasons and if he is in doubt that the things are going in right direction (endangering safety!), he may propose the Exceptional and/or Special Audit for such cases. Anyway, this is not a situation how to behave with the new companies which need to be approved for the first time.

Regarding the Audit Schedule, it will depend on the number of Team Members (auditors) and from the type and the scope of the Audit. For the Approval and Regular Audits, the scope is the overall SMS, but during the On-Site Audit, everything cannot be checked. So, during the Documentation Audit, every document is thoroughly checked, but for On-Site Audit (due to limitation of auditors and time, there is need to do Sampling (explained in Section 8.5!).

One example of an Audit Schedule is given in Table 8.1. Let's explain the main points in the schedule.

---

* E-mail is good for such purposes, if there is no regulation on dealing with this letter in some other way.

TABLE 8.1

Example for Audit Schedule for Engineering Department in ANSP

AUDIT	Day 1			Day 2		
Hour	Team Leader	Auditor 1	Auditor 2	Team Leader	Auditor 1	Auditor 2
8:00–9:00	Opening Meeting			Site visit of MET	Site visit of GP	Control and Monitoring Room
9:00–10:00	Critical and Sensitive Areas	Control and Monitoring Room	Checking ADP of Radar Data			Site visit of Radars
10:00–11:00	Site visit of COM	Site visit of VOR/DME		Checking maintenance records for MET	Site visit of LLZ	
11:00–12:00						
12:00–13:00	Lunch			Lunch		
13:00–14:00	Management Commitment	Checking maintenance records for NAVAIDs	Checking maintenance records for Radars	Preparing Interim Report		
14:00–15:00						
15:00–16:00	Team Meeting			Closing Meeting		

Acronyms used in the table: COM is Communication; VOR is VHF Omnidirectional Range; DME is Distance Measurement Equipment; NAVIDs is Navigational Aids; MET is meteorological equipment; GP is Glide Path; LLZ is Localizer; ADP is Automatic Data Processing. All of them are different systems used in providing Air Navigation Services.

As it can be seen from the schedule, this is a two-day audit on the Engineering Department of a small Air Navigation Service Provider (ANSP) on the small aerodrome. There are three auditors (Team Leader and two auditors) and all of them are doing the audit.

8.4.3 Executing On-Site Audit

On the proposed and accepted date, the Audit Team show up in the company premises in the morning. Somebody will meet them and they will register themselves for entry into the company. If there is a need for Security Passes, this must be done in advance. Anyway, the auditors from the Regulator in Risky Industries must be reliable persons, so getting the Security Passes should not be a problem.

Each On-Site Audit will start with an Opening Meeting. This is a meeting inside the company premises where Audit Team meets with the company representatives. Although the Audit Team are guests in the company, the Audit Team Leader is in charge of conducting the meeting. He introduces himself and the Team Members, confirming that the previous arrangements regarding the audit are not changed. He will explain (again!) what the purpose and scope of the audit in accordance with the Audit Schedule. He politely explains the rules and procedures which will be followed during the audit. If there are changes in the Audit Schedule, they need to be elaborated clearly.

The company, from their side, introduces their representatives. Each auditor should be accompanied by (at least) one company's representative who will provide help to the auditor. These company's representatives are chosen by the company and they are usually employees in particular department (or unit) which is subject of the audit. As such, they are familiar with the rules and procedures in their department, and they take care of all possible requirements from auditors. It is strongly requested for auditors to be accompanied by company representatives also from the point of their protection: They are not familiar with the safety and security rules for accessing particular areas (This is a Risky Industry!!!) and they do not know always where and which Personal Protective Equipment (PPE) shall be used.

Usually the counterpart of the Team Leader is the company's Safety Manager, but maybe company will decide some other (higher) manager will be in charge. Let's say, if they have Compliance Manager, he can be the counterpart. The counterpart will (maybe) provide some notices regarding the Audit Schedule and there is nothing strange with that: The Audit Schedule is a life document and it can be changed reasonably! The meaning of "reasonably" is that the Team Leader will consider the proposed changes by the company's counterpart, and if they do not affect scope and targets of audit (mentioned in Target CL), the changes may be accepted.

The company, at this meeting, will inform the auditors about the company's representatives who will accompany the auditors. They will also provide information regarding the safety and security procedures inside the company premises. If there is need for PPE (Personal Protective Equipment), it is the time for it to be given to the Audit Team. In the case of wearing some sensors (chemical or radiation!), they will be given also to the Audit Team.

On this meeting, also, the company's counterpart will inform the Audit Team regarding the room which they can use during the audit and the company shall provide access to the room only for the Audit Team.

After the Opening Meeting, auditors start with their activities in accordance with the Audit Schedule. You can notice from the Table 8.1, that the auditors are not together during the audit, because each of them has different schedule and they cover different areas.

Lunch is the next time when the Audit Team can be together again. Lunch is usually a working lunch. It is used to interchange some notices, experiences, and comments between Team Members or to clarify some activities or findings.

The last hour of each day is also dedicated to the Team Meeting. This is the time when the Audit Team Members meet each other in the room dedicated to them. They can use this time to consolidate their findings and/or to discuss them with the Team. This is also the time where, at the end of working day, the audit progression is checked (is it going on as per Audit Schedule and Target CL).

After the lunch on the second day, the auditors will go to the room dedicated to them for preparing the Interim Report. This is the time to produce the Final Report which is called "Interim." Although this report is prepared on the last day of audit and it must be also signed by company's representative, it is not the Final Report. This is just an immediate (interim!) report regarding the findings during the audit and it must be presented at the Closing Meeting to the company. At the end of this report, there is a space where the company may put its comments (notices, complains, explanations, etc.) regarding the audit and/or findings.

After going "home," the Audit Team will meet again in the Regulator's offices. They will reconsider the findings and comments (notices, complaints, etc.) from the company and they will produce a Final Report which will be the consolidated "interim" report. The Final Report will have all the findings and in addition, there will be dates for actions which must be taken by the company to establish compliance regarding the findings of non-compliance.

This Final Report must be produced and send to the company not later than two weeks after the last day of the audit, but more details about this will be presented in the Chapter 9 (Conducting the Audit).

8.5 Sampling

Risky Industries are complex companies and most of them are consist of many companies (or departments). There are many processes in each of these companies and their complexity is expressed as a number of operations, number of staff and complexity of their expertise, type of systems, and equipment used, etc. Even the safety regulation dedicated to Risky Industries is taking care for many aspects of their work, So, speaking about the Risky Industries, we are speaking about the functional system with great complexity.

It is clear that such a system cannot be overseen easily. If the operations are complex, their Oversight will be even more complex. The simple reason for that is that the Regulator must take care not to interfere with company operations during oversight activities, so the overall oversight activities will need more complexity.

There is no one way of how to do the Oversight and this expression totally applies to the audits (which are just one of the tools used by Regulators to do Oversight). It must be mentioned here that audit is very good and comprehensive tool and that is the reason that Regulators must put due attention to how to use this tool to get most of the benefits.

The aforementioned complexity of the companies and their operations in Risky Industries are defining the process of auditing which is presented for the audit by *Sampling*. Sampling is a process of checking processes (procedures, operations, records, etc.) in one company during the audits without checking all processes. Instead, each auditor takes just few samples of any process (procedures, operations, records, etc.) and by analyzing them, he gets a conclusion regarding the overall process (procedures, operations, records, etc.) in the company. What is important to understand is that sampling is connected with On-Site Audit. During the Documentation Audit, everything shall be checked.

Sampling is a well-known procedure in statistical analysis: You choose a representative sample from one big set of data (population) and by analyzing it, you estimate characteristics of the overall set (population). If you would like to implement some change in the product of your company and you would like to investigate what the customers think about this change, you provide a survey. Of course, you may not survey every customer (it is expensive and it needs a lot of time!), so you choose a particular number of them and you do a survey of them using statistics.

Axiom 18: Due to the complexity of the job to be done during the On-Site Audit, the auditor uses Sampling to check processes (procedures, operations, records, etc.) in the company!

Sampling is done twice during audit process. Once when the Team Leader decides what to audit – this is done before the On-Site Audit (during planning it). The second time is during the audit – the auditor will not check every piece of data (process, procedures, records, etc.), but he will make samples of which data will be checked. In simple words: There is systematic sampling, there is random sampling, and very often they are combined!

Systematic sampling is when there is a rule on how to do sampling. For example: There is class of 30 students and there are 15 boys and 15 girls. As a sample, I will use the rule to choose three boys and three girls who will represent the class. Why I am doing that? Because there are two subsets (boys and girls) in my set (students) with obviously different characteristics (physical and mental). So, having in mind that the numbers of boys and girls in the class is same, the numbers must be same also in my samples.

Random sampling is when you take randomly samples from big set trying to do it from different places in the set. As a simple example for random sampling, we can use choosing numbers in the Lotto: All numbers are chosen totally randomly, because every one of them have equal chance to show up on the screen.

In our case of the class from above, I decided to choose three boys and three girls (systematic sampling!). Now, I will choose randomly 3 boys from the subset of 15 boys and 3 girls from the subset of 15 girls. It means that each of them will have equal chance to be chosen!

And this is a simple example of the combination of systematic and random sampling: First, I am dividing set on two subsets (15 boys and 15 girls) which is systematic sampling and then I am randomly choosing three boys and three girls (which is random sampling). If I do not apply the systematic sampling to the class in this case, there is a chance randomly to choose six boys or six girls and using these samples, my analysis will not give a correct result regarding the class.

How to do sampling in the sets (population) is a big area in statistics, but there is no need for audit purposes to be expert in this area. I hope that next two sections will provide enough advice on how to do proper sampling for audit purposes.

8.5.1 How to Choose a Sample?

The sample chosen in each case (process, records, etc.) must be representative and the number of samples should be significant!

Representative sampling can be defined as a particular distribution of samples which will give good estimation of the set (population) characteristics. Let's say, in the example with the class from above, the sample of three boys and three girls is a representative sample of the class. Choosing an equal number of boys and girls (because the both subsets (boys and girls) are with same number of elements!), will provide representative *unbiased* analysis of some characteristics of the class where differences between boys and girls could be significant.

The biggest problem in providing representative sampling is *bias*. A simple example of bias is choosing six boys from the above-mentioned class of 15 boys and 15 girls. If we try to provide a survey regarding today's fashion by using only six boys (it means that we are neglecting girls in the class!), the results will be very strange and wrong. Simply the boys may not care for fashion in the way the girls do, so bias in the example with six boys will be expressed as "boys' fashion!"

To clarify bias, we can use the same class from above. Assume that the class has 20 boys and 10 girls out of 30 students. To provide a representative sample of population which will be *unbiased*, we should choose four boys and two girls! Simple: Four boys in the sample will represent 20 boys of population (class) in the same manner as two girls in the sample will represent 10 girls in the population (class). Actually, bias is taking care of same ratio of number of elements in the subset and number of samples from each subset. It can be presented by formulas:

$$Unbiased\ sampling \rightarrow \frac{boys\ in\ the\ subset}{boys\ in\ the\ sample} = \frac{girls\ in\ the\ subset}{girls\ in\ the\ sample}$$

Auditors need to be careful how to choose samples, but following approximately the formula above will provide good results.

But there is another important question: Which number of samples chosen from a population could be significant?

It is a fact that having bigger sample, the auditor will get better results, but there is (again!) a need to have balanced approach. Anyway, the question how to choose right number which will provide good results cannot be answered simply!

In the case of auditing, auditor should decide by himself how big the sample will be. In my humble opinion, choosing 10% of the set of records as a sample could be a good significant and representative sample. But, be careful, because for the auditor on an On-Site Audit, the limitation factor is the time. If the numbers of records which are the subject of audit, are of range of 1000, choosing 10% out of 1000 is 100. Could the auditor analyze all of them in the short time dedicated to On-Site Audit? I have doubts about that...

An important thing to mention here is that the Regulator (or auditor) may not establish a general rule on how to do sampling. Sampling is highly dependent on the size and nature of company, complexity of operations, volume of the activities and staff inside, environment, position in the market, etc. So each company will need a different rule of sampling for the audits. The auditor may establish by himself general approximate rule(s) regarding the systematic sampling of data, but even this one shall be done very carefully. The best way is to decide on-site about the number of samples, having in mind the company. Good results with choosing a representative and significant number of samples come with experience.

In general, an auditor during audits should use combination of systematic and random sampling. He systematically chooses a rule on what to focus on. Let's say: He takes into consideration (systematically) only records recorded from last year, last month, or the last audit, and after that, he randomly chooses five to ten of these records as a sample.

Another important thing regarding the sampling is that the auditor must do the sampling by himself. He can politely ask the auditee representative to choose the samples by himself. In any case, he may not accept the samples offered from auditee! He must always have in mind that there are companies from classes 4 and 5 which are prone to provide their best picture or even to fabricate results. The auditor must be careful in how he will refuse this offer for sampling from company's representative and he must be firm (but polite!) when he asks for his own choice of samples. If the auditee representative insists on his samples, auditor can explain that he must follow a rule on what and how to choose. But he shall never (ever) try to explain what the rule is!

Axiom 19: The auditor is the person who is choosing from where, how many, and which samples will be taken! He must not accept the samples offered by company's representative!

8.5.2 How to Use Sampling in Audits?

As I mentioned in previous section, during audits, the auditors are using systematic sampling and random sampling, but each of them in different areas. How it will be done will be explained in this section.

When the Team Leader establishes the Audit Schedule for On-Site Audit, he uses systematic sampling on what to audit. As simple rules for systematic sampling how to choose what to audit, my recommendation is to use these:

(a) If there is something which is not clear to you from documents during the Documentation Audit, you choose this area.

(b) If there is something suspicious or unrealistic in any area which you have noticed during the Documentation Audit, you choose this area.

(c) If there are previous audits or reported events regarding the safety performance of the company, you choose these areas.

(d) If, in the period between two audits, there was a Safety Case submitted by the company about the change inside, you choose this area.

(e) Any area which was the subject of changed regulation between two audits, could be chosen for the next audit.

(f) Any area which your instincts are telling you needs to be checked, choose it.

(g) If this is a Regular Audit, then choose different areas for audit from the previous audit. Auditing the same thing in two consecutive Regular Audits is not recommended if you are satisfied with the findings of the previous audit.

The Company will receive the Audit Schedule for the On-Site Audit and they will know that not everything will be checked. Anyway, there is nothing wrong there. The only thing which needs to be maintained is that the auditors may not explain or comment on why they made such a choice to the company. This could be an advantage for auditors during the audit, because if they explain or comment, the company will prepare only for these subjects, which will not give a real picture regarding company performance.

During On-Site Audits, we are using random sampling. As mentioned before, the main point with random sampling is to establish the equal probabilities of the sample distributions which will provide unbiased representation of the population (set!).

Having in mind the formula for unbiased sample in the previous section, there is a need to establish approximately the size of the set which is the subject of auditing. It can be some list, some activities (Preventive and/or Corrective Actions), some records, reports, measurements, etc. Whatever the size of set of documents is, samples should be in the range of five to ten. It is wise to look briefly around 20 to 30 samples of the set, but for more detailed checking, the auditor should choose not more than ten.

Nevertheless, even the random sampling is dependent on the "personality" of the company, on the type of audit, and on the situation encountered by the auditor during the audit. In general, for an Approval Audit, the company is new and there is not so much data, so the auditors may choose five samples. For the Regular Audit, the following recommendations are offered:

(a) For companies with assumed "personalities" of Class 3, 4, and 5, the auditor must always take more samples.

(b) For companies with assumed "personalities" of Class 1 and 2, the auditor can take five samples.

(c) If the audited performance of the company in the previous audits was good, the auditor can choose five random samples from the set.

(d) If it was not so good, the auditor can choose ten (or more) in the areas where the bad performance was shown.

(e) If the company has introduced a new product or service in the period between two Regular Audits and they have submitted a Safety Case to the Regulator, then the auditor can choose ten samples in the area where the new product or new service was introduced.

(f) If the company changed something in its operation, organization, or structure and they have submitted a Safety Case to Regulator, then the auditor can choose ten samples in the area where the change was done.

(g) If the set of records (lists, measurements, Preventive and/or Corrective Actions, reports, etc.) is well-ordered and maintained, auditor can take five samples.

(h) If the set of records (lists, measurements, Preventive and/or Corrective Actions, reports, etc.) is chaotic and badly maintained, the auditor can take the samples.

9

Conducting the Audit

9.1 Introduction

Conducting the audit depends on the type of audit. As I have mentioned previously, I will put emphasis on the Approval Audit which is most comprehensive type of audit. Having a good understanding of how this audit is conducted will help with all other audits. The rules on how to do an Approval Audit apply to all other types of audits.

In general, each audit consists of three parts: Data collection, data analysis, and presentation of the results. Data collection involves a series of observations made by the auditor who needs to record these observations. Data analysis is executed on data gathered, and it provides results which are usually later compared with industry standards and good practice. As a last step, the results are combined into the Final Report, presenting conclusions drawn from the data analysis and providing them to the company and to the Regulator in the form of objective evidence.

I will stick in this chapter mostly to data collection and presentation of the results, because data analysis is strictly dependent on the nature and quantity of data from the particular Risky Industry and in general, it is too big topic to be presented in this book. Of course, data analysis implies that the auditors must be capable of analyzing gathered data.

In this chapter, I will focus on practical advice on how to deal with Documentation and OnSite Audits, what to check, what to ask, to what to pay attention, and how to get information to achieve that which is intended.

9.2 How to Do Auditing?

9.2.1 Auditing Documentation

Auditing documentation is done in the Regulator offices. The Regulator has received the soft or hard copy of documentation from the company and the

Team Leader, after initial consideration, will split the documentation out to the Team Members (auditors). They will go through all documents and compare what is in the documentation with the regulation requirements. This is a phase of the audit when auditors get familiar with the company and its "personality" through the documentation.

The auditor uses documentation to gather knowledge on what is the business model followed by the company, and for such a purpose, he must possess an analytical mind. This is the first insight which is gathered through the Documentation Audit. Reading the documentation, the auditor must understand two things:

(a) How the company is conducting business

(b) How the company is dealing with Functional Safety

A good auditor must not only find and understand the answers to these questions, but what is more important, he must ask himself: Do these answers comply with each other? As has been said before: There are no two Management Systems the same, so each Management System must be tailored in accordance with the reality (the business model of the company!).

Let me give you a simple example which I have experienced in my past. I was working for an airline in south-east Asia as Quality Assurance and Safety Manager, and the first thing which I did when I arrived was to check the already implemented Quality and Safety Management Systems. I noticed very strange things: Both Management Systems were too robust for such a small airline. Later, I understood that this company had found somewhere the QMS and SMS from much bigger airline and they just made some small changes and implemented it. If you still do not understand what is the meaning of that, please note that the airline where I was employed had four aircraft and the other airline (from which the QMS and SMS were taken) had 200 aircraft... I spent three months adapting the Manuals for the existing reality and from 80GB of electronic documentation, I produced only 20GB. Although, it was four times less, it was totally applicable documentation satisfying all regulatory requirements.

Beside looking for answers to two questions from above, the auditor must check the presence of procedures dealing with all documentation requested by regulation. How the documentation is produced and how it is controlled are important questions which must be answered during the Documentation Audit.

The main point is that all the documentation is important, but not all documents inside provide the same valuable information regarding the performance of the implemented Management System. In my humble opinion, the auditor must pay attention to the procedures which deal with:

(a) How the hazards are identified. (Check the procedure for hazard identification, check the list of identified hazards, look for inconsistencies, etc.)

(b) How the risks are calculated. (Check the criteria for severity and frequency, check the list of risks and the values for each of them, check for irregularities, etc.)

(c) How the internal safety reporting is provided. (Check the procedures, actions triggered by the reports, etc.)

(d) How the Preventive and Corrective Actions are executed. (Check the procedures, check the monitoring of actions, look for inconsistencies, etc.)

(e) How the safety information is distributed inside and outside the company. (Check the procedure, check for irregularities, etc.)

(f) Is there any uncertified software which is used in operations? (As mentioned in Section 1.2 (General Explanation Regarding the Book), there is a regulatory requirement in all Risky Industry to use only certified software for any operation in the company which could have safety consequences and the auditor must ask for software certificates!)

All these things should be clarified during the Documentation Audit and, do not forget: They must be also checked during the On-Site Audit. During the On-Site Audit, the auditors are looking for proof that what is written in the documentation is implemented and followed in the reality. Checking the company's records (reports, lists, measurements, etc.) on-site should provide data which must be in accordance with the submitted documentation. So, it is wise for each auditor to have scanned versions or soft copies of all documentation in their laptops (tablets or mobile phones). It is normal, during the On-Site Audit, for the auditor to compare documentation with the reality on-site.

9.2.2 Auditing Equipment

As it has been mentioned in Section 2.10 (What Is a Management System?), each Management System consists of Equipment, Humans, and Procedures and when auditing the Management System, the auditor must check these constituents.

Auditing the equipment inside the companies is not a big deal. There are rules on how this equipment shall be built and these rules covers the hardware and software. Auditing software can be done only by checking if the software is certified for use in Risky Industries, but auditing hardware is different. Software provides control and monitoring of operations and the hardware performance, so hardware is checked with regard to the performance. Anyway, it is not a big issue. The level of IT implementation in today's industry provides software control which can provide information on hardware performance in real time, so a printout of the data displayed and stored in the computers can immediately provide a good overview of hardware performance. In general, the requirements for equipment reliability are very strong and control over hardware performance requires high integrity where the warnings and alarms are part of overall functioning.

So, only by few moves of the mouse and few clicks on the keyboard, the company can provide the present situation with the hardware performance and also the history of the last few months (subject to regulation and the frequency of events!)

It is good if the auditor asks for:

(a) Printout of the hardware performance at the moment of audit. (This will prove that at the moment of audit, the hardware is OK. You could imagine what a scandal would be for the Regulator if, in the time of audit, the hardware was not OK and the auditors did not notice it.)

(b) A few printouts of hardware performance at different times from the history. (It will provide data for continuity of service of the equipment.)

(c) A few printouts from the alarm and warning logs from the history. (This will provide information about the handling of past problems with the equipment.)

These printouts can be taken, but most of them will be few pages, so checking them in detail by the auditor can be done later (after the audit for that day is finished). What is important for auditor is to check the tolerances adjusted on these printouts, because this could possibly be wrong. Adjusting wrong tolerances will not trigger an alarm if something bad happens with the equipment and the disaster will not be stopped.

But there is something else which is not always followed by the Regulators and by the companies. In Risky Industries, there are regulatory requirements for a few additional characteristics regarding the equipment used in the processes of manufacturing products or offering services. These roughly can be presented as reliability and integrity. These are two characteristics* which are regulatory requirements in a few Risky Industries and they need to be fulfilled.

Reliability is the probability that for particular time the equipment will provide normal operation. Increasing reliability can be done by doubling the equipment and it is done in many Risky Industries. If the one apparatus is not working, the doubled one will undertake the operation.

The integrity is the level of trust which we can put in our equipment. In other words: Integrity gives information on whether the equipment conducts the normal operation for which it is designed. Integrity is provided by

---

* In aviation there are requirements to calculate and maintain particular levels of Availability, Reliability, Integrity, and Continuity of Service for each type of CNS equipment in the aircraft and on the ground.

measurements and monitoring in real time. If the equipment does not provide normal operation, the monitor will trigger the alarm, switch off equipment and the operator will investigate and fix the problem.

Both these characteristics should show if the equipment is well-maintained, monitored, and controlled and this should be checked by the auditor (if it is part of the regulation!) A lot of companies do not pay attention to these characteristics and they just neglect them because there is need for data-gathering and particular personnel, good in mathematics, statistics, and probability, to calculate them. The reason that the companies behave in such a way is also because the Regulator implicitly neglects this also.

9.2.3 Auditing Employees' Qualifications

In each industry, there is a need for employees to be properly educated and skilled in the operations which they execute. Sometimes companies employ less educated persons and they train them for the operations. Or, the employees who have spent few years in the company are promoted into positions which need a higher level of education. It can bring good benefits for some of the industries, but it is not allowed in Risky Industries. There is a simple reason for that: The operations there are very complex and the knowledge of what is going on and what can happen has primary importance.

In Risky Industries, it is not enough to have employees with the proper education. There is a need to have an employee with particular education, skills, attitude, and experience. Let's give one simple example: The same engineers of electrical engineering can be employed in a nuclear power plant and in a manufacturing industry (manufacturing TV sets, for example), but their position and attitude differ. Whatever the engineer of electrical engineering will do in nuclear power plant considering the consequences is not the same as what he will do in the manufacturing industries: The outcome of every activity in a nuclear power plant has considerable safety consequences compared to the outcome of activity conducted in manufacturing industry.

That is the reason that in Risky Industries there is regulatory requirement for licensing and certification for employees as an activity which will prove that they are capable of doing a job which has safety consequences. There are many examples of licensing and certification: In aviation, the licensing is a must for pilots, cabin crew, aircraft maintenance personnel, and for ATCos, while the certification is required for engineers in ANSP and employees in MET and AIS departments. In the petroleum industry there is also the requirement for licensing of the persons who may search, bore for, and extract petroleum (depending on the State regulation!) In the nuclear industry, the persons and the companies also must be licensed.*

---

* In the nuclear industry, the terms "license," "authorization," and "permit" are considered to be synonymous. See the IAEA Safety Standards, Specific Safety Guide No. SSG-1, Licensing Process for Nuclear Installations.

Licensing must be done (if not regulated in another way!) by the Regulator. The Regulator must establish the body (department) which will take care of the process of licensing of employees in Risky Industries. The simple example for such a body is the department in the Ministry of Internal Affairs* which takes care of obtaining driving licenses of road drivers. As you are familiar with this activity, there are some regulatory requirements for training, theoretical, and practical exam in this area. There is a possibility this function to be transferred to another private entity, but in such a case, this entity should be accredited. Certification may be done by the company itself or by another Certification Body or educational institution (it is also a subject of regulation!).

The auditor should check and compare the education and licenses/certificates with the job description of few positions (sampled by him!) which he finds representative for the some of the operations in the company. It is good if licenses and certificates are checked for each person who is the subject of licensing and certification by the regulation, but this is subject to available time. Anyway, there must be a register of all issued licenses in the Regulator's premises and there must be such a register (with trainings and assessment) for the certificates of employees within the company. The auditor just needs to systematically sample a few positions and check the education, licenses, certificates, and training records provided for the employees in these positions.

This is not so critical for managerial positions, because there is no proper education regarding Managers. In Management positions, there must be guys with a particular attitude and experience and, of course, with the appropriate license! There are some operations in the company which can be critical for safe performance, so every educational background, licensing, or certification for such a position must not be improvised. It means that the person in such a position must have the proper education, license, certificate, and training, and, very importantly, must be a person with considerable experience! Simply, you may not put a person as ATC Manager if he is not licensed with the highest ATCo license.

9.2.4 Auditing Safety Policy†

There is not so much to be concluded from Safety Policy during the On-Site Audit, but there is much to conclude about company's understanding of safety during Documentation Audit.

It is shocking how people employed in Risky Industries have the wrong understanding regarding Safety Policy!

---

* In different States, there are different organizations of the entities which conduct exams and provide administration regarding the driving licenses of the citizens.
† Although this section is about Safety Policy, whatever is said inside totally applies to any other type of Management (Quality, Training, Environmental, etc.) policy!

Honestly speaking, this is applicable also for Quality Policy, Training Police, Health Policy, Environmental Policy, etc. In other words: The essential understanding of each type of policy is missing in the overall industry!*

I had a chance to attend workshop for Training Manager Post-holder with the General Civil Aviation Authority (GCAA) of United Arab Emirates in Abu Dhabi, where insights regarding safety were given to us from a person who was not so good in safety. I was disappointed with the questions asked and answers given and I tried to correct these wrong statements. My explanation about the importance of Safety Policy especially produced a reaction from many in the auditorium. One of them told me that there is no need for employees in the company to be familiar with Safety Policy (!) When I asked him if they have signed Safety Policy on the walls in his offices, he answered positively. And then I had to ask: Why then do you have signed Safety Policy on the walls of your offices and corridors, if this is only for Managers?

After that, everybody was silent – I did not get an answer...

The next unpleasant situation was during a regulation audit with one auditor who was auditing Training and Competency Management System which I have produced. I was frustrated that, although the Training and Competency Manual was approved by that Regulation Body, this auditor had not read it. He asked something and I tried to explain to him that this is included in the Training Policy. To my efforts, he only said: Policy is a High Level document and it has nothing to do here! I immediately shut up, but still I wonder about my reason for shutting up: Was it because I was shocked with the statement or because it is not wise to make your auditor angry by opposing him?

I totally agree that the Safety (Quality, Training, etc.) Policy is a High Level document, but it does not mean that it is dedicated to high-level managers! It does not mean that employees do not need to be familiar with it. It is a High-Level document because it is the Constitution of each Management System which is produced!

Similar to how each law in the State must be in accordance with the State Constitution document, the same is valid for each Management System: Whatever is inside the Safety Management Manual must be in accordance with the Safety Policy!

Let me explain this a little bit in more detail:

As I said in Section 2.12 (Understanding Procedures), the procedures are "material" which builds the Management System. So, when dealing with Operational or System Procedures, Managers are trying to provide procedures for each process (operational, activity, task, etc.) in their company. But the procedures cannot always cover all possible situations. It is common that sometimes your employees will experience a situation (event!) which is not

---

* Please understand that I have good reasons to have such opinion and I do not like to push you to agree with that! Please try to build your own opinion.

covered by any procedure. Simply, the Safety Manager did not believe that such a situation could happen (it never happened anywhere in the world before!) or he assumed that it is so rare and so expensive to handle it that he did not provide a procedure on how to deal with it. And how should the company's employees deal with such a situation if this happens?

Will they ignore this situation?

Not at all! The answer is: They will improvise!

They will improvise, but this should not be just a simple and wild improvisation. It must be in accordance with company's Safety Policy! And to do this, the employees must be familiar with Safety Policy!

As the people in the States of the Balkan Peninsula used to say: "The Devil does not dig holes or do any job! All the time, he thinks how to make bad things!"

And the bad things happen... They happen when you do not expect, in the way which you do not expect, and with the consequences which you do not expect.

The Safety Policy has nothing to do with the situation in the company when everything is OK and that is the reason that it is very much underestimated in today's industry. Bad things happen, but they do not happen very often. We produce, implement, and maintain the SMS and we use it in our lives to prevent bad things from happening and if they happen, to eliminate or mitigate (as much as possible!) the consequences. And these very rare bad events, which we struggle to eliminate or mitigate, are contributing to our misunderstanding regarding a Safety Policy.

So, what is a Safety Policy?

Looking on the internet for a definition, you will find a lot of definitions. Most of them define a Safety Policy as a company statement regarding the Management commitment, responsibilities, and arrangements regarding Safety Management. I am ready to agree with this, but I would not include there the responsibilities. There is a simple reason for that: Safety Policy is a "general statement" and in general, safety is an obligation and responsibility to everyone in the company. Usually in the Manual the safety structure of the company is explained (Safety Manager, Safety Department, Safety Unit, etc.), so the details about particular responsibilities should not be mentioned in the Safety Policy. There should be an additional chapter in the SMS Manual where responsibilities are clarified. Also, I had the chance to read the Safety Policy where it mentions: "... the General Manager, Mr John Smith ...." This is also wrong: If they change the General Manager in next few months or years, they will need to change also the Safety Policy, which is a totally unnecessary activity.

In my humble opinion, it is more important to see in each Safety Policy the general principles of how the company will deal with safety. The Safety Policy could be a document which is only one page and there should be four to six considerably substantial statements which will explain the strategy how the company is dealing with safety.

Axiom 20: A Safety Policy explains the Strategy of how the company is dealing with safety! The tactics can be found in the rest of SMS Manual!

Let me explain a simple example of how to use a Safety Policy:

Different companies in Risky Industries have different Safety Policies. Let's say they provide an order of what to save: In the case of an accident, you first save children, after that, women, after that, men, animals, premises, etc. But this does not apply for the nuclear industry. There, in the case of incident or accident, you do not help people until you shut down the reactor. When you shut down the reactor, then you think about how to prevent radioactivity getting outside and then you focus on injured humans. Having a Safety Policy in nuclear industry to save children first instead of shutting down the reactor may cause the deaths of many children (and women, men, animals, etc.).

But there is another aspect regarding Safety Policy and this aspect is sometimes forgotten: The Safety Policy also applies to the Safety Regulator! The Regulator must also produce a Safety Policy which will establish the State strategy of dealing with safety in the State's Risky Industry.

During the Documentation Audit, the Safety Policy is usually audited by the Team Leader at the beginning, but it must also be considered by other auditors. They must be familiar with the Safety Policy because they are checking everything else in the documentation and this "everything else" (procedures, operations, activities, responsibilities, etc.) shall be in accordance with Safety Policy.

There are the things which auditor must take into consideration when assessing the SMS with regard to the Safety Policy:

(a) Is it providing general information (strategy) on how the Safety Policy will be implemented if something goest wrong? (Can the auditor understand this information?)

(b) Is it understandable to everyone? (Ask about the meaning of particular statements there if you find them confusing!)

(c) Is it providing a commitment to deal with safety in the company? (This is maybe strange to anybody, but it has very strong legal reason to be there!)

(d) Is the Safety Policy (as a strategy!) implemented in the company activities (on a tactical level)? (Are the procedures inside the SMS Manual in accordance with the Safety Policy?)

(e) Is the Safety Policy available to every employee in the company? (The standard solution is to have the Safety Policy document on the wall of every office in the company.) It is good to check if the employees are familiar with the Safety Policy, but the results could be very disappointing.

I do believe that by looking for these few things mentioned above, the auditor could gather a necessary understanding about company's Safety Policy. But if there is more data in the Safety Policy, there is nothing wrong with that! I will repeat again: The Safety Policy is written bearing in mind the company understanding of actions which will improve the safety of their products or services offered (in the scope of Functional Safety!)

The auditor may not agree with what is inside the Safety Policy, but he will need objective evidence if he would like to state that something is wrong with a company's Safety Policy. In general, the auditor cannot be sure that the strategy mentioned in Safety Policy is not working, so (in general!) he should not find non-compliance regarding Safety Policy if the few items listed previously are part of the document!

The auditor must be careful when he is checking the Safety Policy! It can provide a lot of data regarding the "personality" of the company, but at the same time, comments given by the auditor to the company regarding the Safety Policy can provide a lot of data regarding the "personality" of Regulator.

9.2.5 Auditing Safety Objectives

This is something which is done during the Documentation Audit and it is important for the Approval Audit.

Very often Safety Objectives are merged with the Safety Policy. There is nothing wrong with that, but I do not recommend it. The reason is that the Safety Policy is a company strategy which is covering a long period and Safety Objectives are tactical goals and aspirations which need to be achieved during shorter period (usually one year).*

Safety Objectives are statements which are presented as goals (targets) to be achieved for particular period of time. The Safety Objectives' importance is big, because they are used to establish the safety performance of the implemented company's SMS. Safety Objectives for particular period of time are compared by company performance and if the Safety Objectives are fulfilled, then the SMS is achieving its goal (improving safety!).

It is important that they should be given quantitatively (as numbers) and not qualitatively (as text or description). A good example I can mention: Company will decrease number of incidents in next year by 10% or company will improve safety events reporting for 20% in next one year, etc.

The reason that it is important for them to be expressed quantitatively (by numbers) is that they need to be measurable and comparable. If you cannot express something by a number, you cannot measure it, and you cannot compare it. Using the two examples from above, please think would it make

---

* There are Safety Objectives that are implemented on the State level and they can be part of the State strategy. They are usually applicable for longer period (three to five years). But company Safety Objectives usually cover a period of one year.

sense to you if the Safety Objective is written as: Company will decrease number of incidents in next year or Company will improve safety events reporting in next one year? Without numbers, there is no sense: These are not Safety Objectives, but these are Safety Wishes...

What will happen if the company is producing Safety Objectives qualitatively (by explanation, not number)?

Different Risky Industries have different regulations in this area.

If there is a regulation requirement which is looking for a quantitative expression of the Safety Objectives, then it must be followed. The company which expresses them qualitatively will have non-compliance in this area. But if there is no such a regulatory requirement, then there is no non-compliance.

Expressing the Safety Objectives by expression (qualitatively), not by number (quantitatively) could very often happen with the companies from Class 4 and 5 (sometimes even Class 3) from the "personality" table in Section 5.2 of this book.

The things regarding the Safety Objectives which the auditor must take into consideration during the audit are:

(a) Are the Safety Objectives part of the SMS Manual? (If you cannot find them, you may ask for them!)

(b) Are they quantitative or qualitative? (If they are qualitative, ask how they are measured and how they are compared!)

(c) Are the Safety Objectives fulfilled for the previous year? (Ask for the report to the Top Management! There is an obligation in each safety regulation to provide a report to the Top Management at least once per year. Safety Objective fulfillment must be part of this report!)

(d) If they are not fulfilled, there is a need for analysis of why it did not happen. (Is there any analysis regarding level of fulfillment of Safety Objectives?)

In general, auditing the objectives should not be problem for the auditor, because if the objectives are quantitatively expressed, then it is easy to understand them and it will not take a lot of time. Auditing of Safety Objectives is usually done by Team Leader.

Regarding non-compliance, usually the audit cannot show any non-compliance regarding the Safety Objectives. If the Safety Objectives were not satisfied in the previous year, it is a company problem. But this is information which should trigger the Audit Team to investigate what was the reason for non-fulfillment. This information could be significant in determining the "personality" of the company. Auditors must be careful with the Safety Objectives. There is not always something significant about the safety performance of the company hidden behind non-fulfillment of the Safety Objectives. Findings in this area should be expressed only as Level 3 findings (Observations), but this could be based on "case-by-case" scenario.

Also, the auditors must understand that the Safety Objectives are a life document and they can be changeable, simply because they are usually covering period of one year. Next year the company will need to produce new Safety Objectives and these need to be send to Regulator as information, not for approval. There is nothing which needs to be approved regarding the Safety Objectives.

9.2.6 Auditing Procedures

I will repeat the Axiom 4 from Section 2.12 here again:

Axiom 4: When the auditor is auditing the Management System, he is checking the existence, implementation, performance, and maintenance of System Procedures.

I just like to emphasize what is written in this Axiom: When auditing SMS, you are auditing company's System Procedures! There is a question here: What about Operational Procedures? As a reminder, these are procedures which explain how humans are using equipment to produce products or to provide services. These procedures shall be checked for safety also, but there is strange mixture between Occupational Health and Safety (OHS) and Functional Safety here. Anyway, that what the auditor checks at the beginning is the list of identified hazards! If these hazards are connected with any Operational Procedure, you may decide to check it. I said "you may," but you need to take care because, as explained in Section 8.5 (Sampling), you do not have time to check everything during On-Site Audit.

I was pretty active in the Safety Groups of LinkedIn in the past. I have investigated a lot of quality and safety topics there (some of them very provocative!) and I have gathered a very useful amount of information there. In one of my posts I received unexpected number of frustrated complaints regarding the procedures implemented in Quality Management Systems (QMSs) and SMS. In that time, this frustration was unexpected to me, but progressing with my quality and safety knowledge, it becomes understandable: There is really a lot of wrongdoing regarding the procedures in all industry.

So, what are procedures?

Procedures can be defined in a lot of different ways.

I am defining them as written instructions (documents!) produced by companies where employees can find information (and command!) on What, Where, When, by Whom, and How a particular job (activity, task, operation, etc.) can be done.

And this definition can be used by the auditor to audit procedures! He just needs to read the procedure and if, during the reading, it is clear to him What, Where, When, by Whom, and How the job (activity, task, operation, etc.) can be done, the procedure is OK.

Here, there is a problem…

The problem is that the procedures are not always expressed by textual explanation which can provide an answer to the questions What, Where, When, by Whom, and How. Very often the procedures can be produced as a Flow Chart and unfortunately, here these questions cannot provide a satisfying answer. Flow Charts are generally used to explain processes (and they are a really good tool for such a purpose), but companies started to use them also for procedures.

A Flow Chart in combination with text and pictures and diagrams is good thing to produce any type of procedures, but a Flow Chart alone is too rudimentary a document for procedures in Risky Industries. In Risky Industries, safety must be provided without any uncertainty and Flow Charts are very much the subject of wrong understanding. I strongly recommend that the Regulator put in a regulation requirement that procedure may not be produced as Flow Charts only.

Good procedure should have text and pictures (One picture – Thousand Words!) If you do not know how the good procedure should look, open any user manual for mobile phone or for car. You can see there step-by-step explanations supported by pictures with embedded numbers inside (for each step!). Looking that, there is no need for high education to operate the mobile phone or car…

But can the procedure which does not provide answer to above-mentioned questions be used as reason for non-compliance?

I feel sorry, but it cannot be used!

If the procedure is missing an answer to some of the questions from previously, it does not mean that there is non-compliance … The reason for that is that what is important regarding the procedures is their performance. Nevertheless, the procedure is incomplete; a company can always explain that they did not experience any problem applying this procedure, and there was training for this procedure where the missing things were explained on the training. You will say: I will look for evidence that it was mentioned during the training (it must be documented also!), but I would not advise going in that direction. It will make the audit idle, bureaucratic, and very time-consuming without any benefit.

If you are not satisfied with the procedure, you may speak to employees of the company (who are using this procedure) and you will record their opinion. If they are aware about all possible things which can go wrong and they feel that procedure is giving them enough information how to do their job, then the procedure is acceptable.

But, if you find an employee(s) who is(are) not satisfied or he(they) is(are) confused by the procedure, you may raise an Observation. It will probably trigger some activities from the Class 1 and 2 (maybe 3 also) companies, but that is everything which can be done.

The very important aspect of the procedure is that employees must be provided with formal training regarding procedure. Formal training means proper training with syllabus, plan, attendance, presentations, material, etc.

should be provided. This can be easily checked by looking at the Training Records. The problem is that not all Risky Industries are looking at their regulation for particular formal training, so companies are free to choose what type of training they can provide to their employees.

There are three big issues with the training regarding the procedures:

(1) Very often the training is too formal and too bureaucratic (especially for Class 3, 4, and 5 companies!).

(2) There is nothing in the records for provided trainings regarding the possible mistakes/errors during following procedures. (This should be provided as Recovery Training: How to recover an operation if something is going wrong or the procedure is not working?)

(3) There is nothing in the provided training regarding consequences which can arise if the procedure is not followed. (Providing such a training is important because most of the employees do not know anything about consequences which could happen if the procedure is not followed!)

The most critical are issues 2 and 3! In the Risky Industries, every mistake/error can produce terrible consequences and that is the reason that the possible mistakes/errors during procedure execution must be explained together with the ways to recover an operation or to eliminate or mitigate the consequences. This is part of Proactive Safety (Think in advance what can go wrong!) and company should prove their proactive attitude by providing training for these mistakes/errors and for consequences.

The issue with the consequences if the procedure is not followed is similar. If these consequences are explained to the employees, they will be aware about danger which is waiting them if they do not follow procedure. And believe me, they will think at least twice, before they decide to avoid the procedure.

My experience tells me that issues 2 and 3 are very common today and they shall be covered by regulation. Regulators must take care about that!

So how to check procedures?

By auditing the SMS, the auditor checks the procedures and he pays attention to:

(a) Is there any missing procedure? (During the Documentation Audit, the auditor checks for the presence of all procedures which cover all requirements and processes in the company!)

(b) Are all procedures (stated in the SMS Manual and audited during Documentation Audit) implemented in the company? (The auditor chooses a few critical procedures (systematic sampling!) and he will check their implementation in the company during the On-Site Audit!)

(c) Are the employees trained for the procedures? (The auditor chooses a few training records for procedures and check the attendance and content of the trainings!)

(d) Are the procedures followed by the employees? (The auditor chooses a few working positions and check if the procedure is followed!)

The auditor must be very careful during On-Site Audits regarding the procedures! You may not like the way the company is writing and dealing with procedures, but you need to prove that some safety event happened due to wrong procedure. It is called Objective Evidence!

It is not just important to have a procedure for each particular process, but the procedure must be effective. You do not check efficiency of the procedure, but you may check if the efficiency can affect the effectiveness of the procedure. For example: If there are too many resources included into the procedure execution, it means that it is a complex procedure, so it may not always provide good results. Check the records from the process covered by this procedure and you can see how things are going on. If there is doubt about anything, then look for objective evidence that complexity is affecting effectiveness!

Anyway, only missing, not implemented, or not used procedure will provide a finding of Level 1 or Level 2. Most of the problems with procedures will provide Level 3 finding (Observation). A good Regulator will provide the regulation regarding procedures and it will provide good training to the companies where he will explain all regulation and all necessary things which are required to produce good procedures.

The ways the procedures are produced and the style and the content of the procedures give very good picture about the "personality" of the company.

9.2.7 Auditing Records

As it was explained in Section 2.12 (Records), the records are actually proof that some procedure (task, activity, operation, etc.) has been followed or some rule is abided by. All these records regarding the procedures (tasks, activities, etc.) or the rules are mentioned somewhere in submitted documentation from the company (usually they are part of the company's Manual). The company must maintain records to prove to itself and to the Regulator that things are going on as planned. The important thing to understand is that the ways the records are produced and how they are maintained must be part of the Operational and/or System Procedures. How and where they are stored, who is the owner, how they are maintained, and who is analyzing them and how, etc. should be part of the particular procedure.

The sampling process is very important with auditing the records, so the auditor must pay attention to how he will choose the records by sampling. In general, he is checking the quantity and quality of the samples as a proof of compliance or non-compliance of overall SMS.

The first thing regarding the records which the auditor must check is the procedure(s) for creating and storing the records. It may not be one procedure, simply because different types of records may have different places to be created and to be stored. It is enough to check just few of them by sampling. Usually these procedures are similar and mostly they can differ by place of storing and time for keeping them. This is valid for paper documents, but if the records are stored electronically, there should be different folder(s) for each type of records. Auditing records is mostly about how the company produces, controls, retrieves, recovers,* secures, changes, owns, and destroys the records. Anyway, the integrity of records should be checked in all levels and in all areas.

One of the important aspects of the records is how they are secured, but this affects security more than safety. Anyway, the auditor must check the security of the records during audits.

Having a list of stored records shows that the company has control over records, but the list and the stored records shall be in accordance. If there is no list requested in documentation and company does not have it, it does not mean non-compliance. But if the list of records is required by regulation, the company must provide the list of records.

It very often happens that some of the records from the list cannot be found in the storage area and vice versa: There is a record, but it is not mentioned in the list. This non-agreement is something which is always good to check. If the non-agreement exists, the auditor can briefly calculate the volume of this non-agreement. If this is less than 5%, it is acceptable, but the Level 3 finding (Observation!) must be mentioned in the Final Report. But on the next Regular Audit, the auditor must check this non-agreement again and if it is repeated, it could produce a Level 2 Finding (Minor). It should increase the seriousness of the company regarding the records.

In each Risky Industry, there is a regulatory requirement to secure the records, so this is also the area where the auditor must check. Security differs from safety, but there are some records which, if the security is breached, can endanger safety of operations.

Records can be written (typed) paper documents or electronic files, which are extensively used today. There are a lot of software programs which deal with records and a lot of companies use these programs. These are generally known as Audit Trails, and they can be used by a company to prove to auditors that they are following procedures and rules from the SMS Manual.

Audit Trails are commercial software programs which allow to the user to electronically maintain the records of sequential events as a tool for supporting documentation (history!) that is used to operations (tasks, activities,

---

* In the nuclear industry, there is a regulatory requirement for the recovery of electronic records after any event that causes disruption to the storage system or a complete loss of a records due to any reason. For more details, you can see ASME NQA-1: 2015, Part 1 (Quality Assurance Requirements for Nuclear Facility Applications).

etc.). The records are time-stamped and there is useful information (details) inside associated with the procedure (task, activity, operation, etc.).

It is easy for an auditor to check Audit Trails, because it can be done from one computer for all company and sampling can be done very easily. Anyway, in the case of any type of records (paper documents or electronic files), the auditor must check a few of them and he must also print a few of them as a proof of compliance or non-compliance.

Records must be stored for particular period of time. In addition, they must be secured from unauthorized access and from unintentional or intentional destroying. In the case of electronic records, having a RAID disk in the computer system for protection of the records in the company is a good thing.

The auditor must check records which are in agreement with the prescribed period of keeping records. This period can be determined by regulation or (if there is no regulation) by the company. He may not ask for records older than this prescribed period, but if the company is still keeping them and they show them to the auditor, then he may ask to check them. Anyway, the auditor must keep in mind that he is the auditing the present situation and the old records must be audited with due understanding that they happened in the past.

There is possibility that old (but still valid) records show deficiency, but auditor may not jump into fast conclusions. If the old records show that the company was bad in some area in the past, and now the present records show good performance, it means that they have improved them. The auditor could take a note and check the same records in the next year's audit. It will show to him if it was just in the past or if this is something which systematically (and maybe periodically) repeats.

If there is a prescribed period for keeping records, the auditor must clarify how they are destroyed and he needs to check the procedure for doing it. This is done during the Documentation Audit, but during the On-Site Audit, he checks the implementation of the same procedure. In addition, the records about destroying of records must be kept by the company. If there is a missing record and it was destroyed, the company could provide proof that they are working in accordance with the regulation.

Anyway, the auditor must have a proper understanding about the records in each company. It is not good if the records are missing, but it is not terrible if the old records are not destroyed. Having more records will not endanger safety, but the auditor must be careful: Significance of nondestroyed records strongly depends from the answer of question: Why are the records used?

If they are used only as proof that the implemented SMS is operational and maintained well, then old records (which are not destroyed!) cannot produce any harm. But if the records are used in calculation of the risks or assessment of the safety performance, then the auditor must be careful: Using old (outdated!) records can produce a wrong picture for calculated risks or for safety performance of the company.

9.2.8 Auditing Preventive and Corrective Maintenance

There is a requirement for the companies in every industry to provide preventive and corrective maintenance for their equipment used in the processes of manufacturing products and offering services.

Preventive maintenance is maintenance which is executed when the equipment is OK. Reasons for that type of maintenance are to change the parts which are subject to wearing and to do cleaning of equipment with the intention of preventing equipment fault. Equipment faults usually trigger failure of operation, so it is important equipment be maintained. The name "preventive maintenance" is coming from the fact that this type of maintenance should prevent faults. For example, when I am bringing my car to the workshop every 10,000 km or every year (whatever happens first), the employees there change the oil (which is dirty from the gases of burned fuel), they change the filters (which are dirty from the impurities in the fuel and the air), and do some checking and readjustments, if necessary. My car is OK before I bring it to the workshop for preventive maintenance, but after this (annual) visit to the workshop, the probability that my car will not fail will increase.

Preventive maintenance is done in time periods and in the way which is recommended by the manufacturers of equipment. The time periods may be changed if the company notices that the equipment is doing well (increase the periods) or doing bad (decrease the periods), but the reasons for change must be documented with particular analysis. Simple example for that can be (again) my car: If I am not driving it very often, of course, then my oil and my filters will not be so dirty and I can go every 18 months to the workshop instead every year. In aviation, there is need for flight calibration* for navigational equipment and radars approximately once per year, but if the calibration results show stable equipment (negligible changes of performance during this year!) than it can be done every 18 months or once in two years.

Corrective maintenance is done when something is wrong with equipment or with the operation. It is usually a process of fault-fixing for equipment and failure rectification with the operation. Preventive maintenance is regular and routine, but corrective maintenance happens not very often. So, it requires more efforts from employees in the company. They must have particularly good knowledge about normal functioning of the equipment and must be familiar with the methods how to find what is wrong and how it can be fixed. Additional skills are also necessary for corrective maintenance. Corrective maintenance in industry is an activity similar to going to the doctor when you feel sick: The doctor will diagnose the problem (find what is

---

* Calibration of the sensors, measurement systems, and instruments is part of preventive maintenance and must be treated as such! Some Risky Industries even require calibration of measurement systems to be done by a company which is accredited by accrediting bodies recognized by the International Laboratory Accreditation Cooperation (ILAC) Mutual Recognition Arrangement (MRA). In some industry, it is enough that the calibration company is accredited in accordance with the standard ISO/IEC 17025.

wrong!) and he will provide a cure (fix the problem and recover your body functioning).

The auditor must pay attention to these two types of maintenance, simply because these are important for proper functioning of the company's operation and if not maintained, they can cause problems with safety consequences. The important things to check during the audit are similar to those for auditing procedures (Section 9.2.6):

(a) Check the procedures for preventive and corrective maintenance! (Both types of procedure shall provide answers to the questions: What, Where, When, by Whom and How. Check are the responsibilities clearly defined for each procedure, etc.)

(b) Check the implementation of the maintenance procedures in reality! (Are they implemented, are the employees trained for them, do the employees follow the procedures during maintenance, records, reports, follow-up actions, etc.).

(c) Check the frequency (time periods) for preventive maintenance for particular pieces of equipment! (Choose a few samples and see do they follow manufacturing recommendation regarding periods for preventive maintenance.)

(d) Check the reports from preventive maintenance! (Their regularity and how much time they lasted.)

(e) Check the reports from corrective maintenance! (How often it happened, how much time was needed to fix the problem – MTTR* shall be less than 30 minutes.)

(f) Check is there any corrective maintenance event triggered by finding during preventive maintenance. (It can show that preventive maintenance is conducted by due attention!)

(g) Check the Follow-Up actions after the execution of corrective maintenance! (Very often there should be improvement of the situation triggered by the fault which was fixed by corrective maintenance!)

(h) Check the values of reliability (MTBF[†]), integrity, continuity of service, etc. (In a few Risky Industries, there is a regulatory requirement to maintain all these operational specifications. Reliability (expressed by MTBF) should not be less than 5,000 hours.[‡])

---

* MTTR stands for Mean Time To Repair. It is the average time needed to fix the problems with the equipment.

[†] MTBF stands for Mean Time Between Failure (Faults). It is the measurement of reliability of equipment. It is expressed as hours and it shows the probability that time between two consecutive faults of equipment will not be less than the specified one.

[‡] This is not a regulatory requirement. It is just a number gathered through my experience that the reliability of the equipment in the Risky Industries (expressed by MTBF) must not be less than this number. Different industries have determined different numbers for MTBF and they could be bigger or smaller than this one.

(i) Check the implementation of the calibration procedures and check the lists for calibration of measurement systems used in company's operations. (Sample a few of them and check periodicity and integrity of data inside.)

Putting particular emphasis on auditing this area of a company's operations is with good reason, because preventive maintenance is preventing faults and corrective maintenance is fixing faults. Faults of equipment and failures of operations are problems and problems behave like children: They grow up with time and one day they will become bigger than you and you cannot change anything (cannot fix the problems!)

9.2.9 Auditing Preventive and Corrective Actions

Preventive actions are actions executed with the intention of eliminating or mitigating a risk which was identified during normal operations and which the executed analysis showed can materialize. This also shows the capability of the company to monitor their operations and to implement a proactive safety approach based on data gathered during daily operations. All daily (or monthly) gathered data about the performance of the SMS, should be regularly analyzed, trends should be determined, and if there is a need to adjust the risks (eliminate or mitigate), the preventive action must be executed.

Elimination (mitigation) of the risk can be achieved by decreasing the level of severity and/or decreasing the level of frequency of the occurrence of the incident or accident. Total safety does not exist and in any Risky Industry, there is a statement of the probability of a particular risk materializing* which is acceptable. This is usually expressed as a number which presents the probability (chance) of the risk materializing.

Corrective actions are actions which could be divided in two categories. One category is action(s) undertaken to fix a particular problem in the company which can be the possible cause of an incident and/or an accident in the future or to fix the non-compliance in the Management System. It is common that the corrective actions of such a type can be triggered by findings during preventive actions.

The second category is action undertaken when incident or accident has happened and the consequences must be eliminated or mitigated.

Both of these actions (Preventive and Corrective) are part of the area which I call Applied Safety, because these actions are actually improving safety.

When auditing them, the auditor must:

---

* For aviation, the probability of accident happening must be better than 1.55×10^{-8} flight-hours. For the nuclear industry (I read somewhere) that the probability of a major accident happen should be less than 2.86×10^{-4} per year (once in 3,500 years).

(a) Check the procedures for these actions. (Check the procedures (What, Where, When, by Whom, and How) and their sustainability, are the responsibilities clearly defined, etc.)

(b) Check the list of preventive actions undertaken in the past. (The number may not be big, but the lack of any such Actions can erase a picture of perfect SMS which rarely exists.)

(c) Sample a few of the preventive actions and check how they were executed. (Look for their effectiveness, for illogical actions, or for strange schedules of events.)

(d) Check the list of Corrective Actions undertaken in the past. (Maybe an incident or accident has not happened, but there could be some observation or adjustments of the SMS. Again, having no action could erase a picture of perfect SMS.)

(e) Sample a few of the Corrective Actions and check how they were executed. (Look for their effectiveness, for illogical actions, or for strange schedules of events.)

(f) Check are the procedures for Preventive and Corrective Actions followed during a few executions of the actions. (Following of the procedure must be maintained!)

(g) Check the follow-up monitoring of a few of these actions. (If there is any, they must be monitored for a particular period of time just to prove that they are effective.)

Preventive and Corrective Actions are important indicators of the capability of the company to handle safety issues.

9.2.10 Auditing Management of Change

Management of Change is part of the business. Very often companies must adapt to new regulations or a new situation in the market, so they are trying to be fast in these adaptations. Adaptations mean that they need to change something in their structure, organization, or technology. The main point with these changes is that they always affect safety.

In this area, there are two terms which are used: Modification and Change. Both of them are used in most of the world's languages, but they differ. For the purpose of this book the "Change" is a change of the operation (activity, procedure, equipment, software, facility, employee, etc.) with the new one. So, the old thing is not in use anymore and the new thing is in use. Modification is also change, but it effects change to just a part of the operation (activity, procedure, equipment, software code/program, facility, employee, etc.) with the new one. Modification can also be just a simple readjustment of the present equipment or data.

In both cases, the change is triggered by need for solving some problem or improving performance. Whatever this "new thing" is, for the "Change," the

Safety Case must be produced by the company and for "Modification," the production of Safety Case is subject to the volume of modification. "Volume of Modification" can be defined as how much the Modification will affect the operation and the employees. Anyway, the company is responsible for its safety performance, so doing an analysis of the Change is under their responsibility.

The Management of Changes is actually a regulatory requirement in Risky Industries. There is a regulatory requirement for the SMS that every change shall be the subject of a new Hazard Identification and risk evaluation. Although there is an SMS established in the company and all hazards are identified and risks calculated, the change in organization, structure, or technology will introduce new hazards and new risks or they will change some of the existing hazards and risks. The procedure for Management of Change is something which needs to be part of the company's SMS.

The point which is almost always forgotten is how the employees accept the change. Humans are, in general, reluctant to change, so it is important to deal with the process of explaining to the employees a few important items:

(a) Why the change will be introduced.
(b) What will be changed and how?
(c) What are the benefits from the change?
(d) What are the risks with the change?
(e) What are the remedies if some of the risks materialize?

There is a need to manage the changes, even the change is triggered by Preventive or Corrective Actions, Usually, a company in a situation when they have decided to make a change, will inform in advance the Regulator with a document called a Safety Case.* The Regulator analyzes the document, checks the effectiveness, efficiency, and sustainability, and he can approve it or disapprove it. This is usually done based on submitted documentation. Some of the changes could require a regulatory On-Site Audit and later, the Follow-Up Audit may be organized in such a situation.

During Regular Audits (which happen once per year), the auditor has to pay attention to:

(a) Checking do they have control over the Safety Cases submitted to the Regulator. (There must be an updated list somewhere or there must be folder (drawer) with documents.)
(b) Sampling a few changes and checking are they effective and sustainable. (Introducing the change is the subject of follow-up activities by the company, checking the measurements and documentation regarding these follow-up activities, checking the results

---

* See Section 2.14 (Safety Case) in this book.

from the activities, and checking the corrections and adjustments, if necessary.)

(c) Check if the procedures in connection with these changes are updated. (There must be Operational and/or System Procedures which are affected by the changes.)

(d) Check were the staff informed about changes or was the training executed. (The training may be planned in the Safety Case, but in general, every change must be the subject of training or briefing* which will provide information from the important items mentioned above!)

The most important and most often neglected is item (d)! This sharing of information regarding the change in the form of training or briefing must provide enough information to employees and it will help the change to be successful and sustainable.

Axiom 21: How the company is dealing with the changes in their operations is a good indicator of how the company is dealing with implemented SMS in total!

9.2.11 Auditing Alarm Systems

In each Risky Industry, there are many tools or devices which provide warnings and alarms. When something is not going as planned, there are Alarm Systems based on sensors and measurement equipment which give warning of an unsafe situation. These systems and devices for unsafe situations can be fire (heat) sensors, chemical detectors, radioactive sensors, current and voltage measurement devices, etc. Mostly these Alarm Systems are measurement systems with adjustable tolerances. If the value of the measured subject (usually something dangerous!) is within the tolerances, then everything is OK. But if the value exceeds the tolerances, that warning/alarm is triggered (siren or light or both!), everybody must leave the premises, and Emergency Procedures are automatically executed (by humans or by safety equipment).

There is a difference between a warning and alarm.

A warning is triggered by the Alarm System when the value of the measurand† is approaching tolerances. The reason to have a warning is to warn the employees that the value of the measurand could produce an uncontrollable process which can endanger the humans, assets, and/or the

---

* A briefing is a short form of information provided to employees regarding some issue, some activity, or some change. In this case (change in Risky Industries), it must be provided only for small changes. Training is the primary tool by which to inform the employees regarding the change.

† Measurand in Metrology (science for measurements) is defined as the value of the system which is the subject of measurement process. For our purposes, a measurand is something which can endanger the lives of Humans, assets, or the environment (heat, poisons, radioactivity, concentration of gases or liquids, etc.)

environment. Undertaking action when a warning is triggered could stop the process reaching the unwanted dangerous level.

An alarm is triggered by the Alarm System when the value of the measurand has passed the tolerances and it is not safe anymore to conduct the process (operation, activity, etc.). An alarm usually gives information to immediately implement Emergency Procedures and to leave the premises.

It is understandable that the warnings are adjusted lower than the alarms. For example: If the dangerous temperature for some liquid will trigger alarm at 60°C, then the warning can be adjusted to 55°C. If, due to some reasons, the temperature of the liquid reaches 55°C, the warning will be triggered and employees would check the reason for this temperature rising before it reaches 60°C. This warning will prevent a safety event if maintained in the proper way.

So, important things to check during Regular Audits must be the functioning and maintenance of the Alarm Systems and calibration of the sensors and instruments. The auditor should check during Documentation and On-Site Audits:

(a) Is (are) there procedure(s) for monitoring, control, maintenance, and calibration of Alarm Systems? (If there is (are) such a procedure(s), the auditor must check are they reasonable and sustainable, do these procedures provide protection regarding safety events, who is responsible for their execution, etc.)

(b) Check are the employees trained for these procedures. (There must be training records, check the content of training and check attendance, everybody must be trained with regard to these procedures!)

(c) Check the implementation of these procedures in reality. (Are these procedures followed during daily activities? There must be some records regarding warnings and alarms.)

(d) Check (especially!) the maintenance and the calibration records for sensors and instruments of Alarm Systems. (If not calibrated, they cannot provide the requested information!)

(e) Check reports on incidents and notice who reported and why these safety events were reported. (There must be involvement of the particular Alarm Systems into these happenings. If not, or if the level of alarms by the systems is too low, it needs to be checked by the auditor).

In my humble experience in this area, the critical issue is calibration* of sensors and measurement instrumentation. In the company's documentation,

---

* Calibration of measurement equipment used during production process is part of legal requirements in many countries! This the way States protect the public from products or services with wrong values! The calibration of sensors and instrumentation of Alarm Systems could be part of these regulatory requirements.

there must be data about the period of calibration for these devices. Anyway, the auditor must understand that the calibration is time- and money-consuming, so it is a burden for companies. Very often, the company tries to prolong the period for calibration and instead of following the manufacturers' recommendation, they establish their own calibration period which is considerably longer (instead of having calibration each year, they do it every two years!) So, the auditor must check the calibration records!

Maybe there is a good reason for having a different calibration period from that recommended by the manufacturing company. In such a case, some of the companies will provide results of previous calibration saying that changes of sensors and instrumentation were negligible for the established calibration period, so they decide to prolong the period. But if this happen, the company must have the documentation (procedures for calibration!) updated for this issue.

Not having calibrated sensors and instrumentation of Alarm Systems is a Level 1 (Major!) finding!

9.2.12 Auditing "Chronology of Events"

Under the name "Chronology of Events," I am considering a method used in audits which connects different operations, tasks, or activities. This is a method which is gives a very good picture how the interrelating of the processes in the company works. The Management System is about managing humans and processes in the company and there is huge correlation between them. So, following one activity will guide you to another activity, this one will guide you to another, and there is a chronological sequence of events which must finish with a good result (product or service).

Let's give an example:

In the industry, everywhere, there is a process of solving problems which show up. For example, there is an equipment fault which can create a failure of operation and if the consequences of this are safety-related, then this is a safety issue. In the SMS, the company will establish a procedure which will manage the process of registering and reporting such an event and the activities which need to be undertaken to resolve normal operation. In general, there will be a procedure for monitoring (How the problem (fault/failure) will be noticed), a procedure for reporting (To whom the problem needs to be reported), and a procedure on how to deal with the problem (Who needs to fix the problem?)

The main point is that all these events come in a chronological sequence and if everything is OK with the SMS, all the history of events for solving the problem will be shown in the records. For example: The auditor must check the times when the problem was reported, when the Corrective Action was undertaken and what is written in the report after the problem was fixed. It is clear that these times will differ in time chronologically: First, the problem shows up, then there was a report delivered to the Maintenance Team, the

Maintenance Team fixes the problem, the Report about Problem-Fixing was written, and the follow-up is the last.

So, the auditor may choose one fault of equipment (or failure of operation) and check (chronologically!) how it is solved. The auditor should check records about:

(a) Registration (track!) for the fault. (There should be a record in the Equipment Monitoring Desk about the fault!)

(b) Reporting regarding the fault. (There should be a record in the Maintenance Room, because they will be informed so they can solve the fault!)

(c) Activities undertaken to fix the fault. (There should be a record of the maintenance staff after fixing the fault! (Report what was the problem and what has been done.))

(d) Material and spare parts used. (There should be a record in the Store Room for the materials or the spares used to fix the problem!)

(e) Post-fault activity in the company. (There should be a record (again!) in the Equipment Monitoring Desk of how the equipment behaves after it is fixed (Follow-Up Activities!))

The main point is that in each of these records, there must be a timestamp on them and the auditor should check the chronology of all these events. The times and dates must agree with each other and it must be a reasonable agreement. If they do not agree, then it does not mean that procedures were not followed, but (anyway), it must be checked why the times and dates do not agree. The auditor should be careful if there are erased and rewritten letters or phrases, because it could mean fabrication or adjustment.

This is a very powerful method which gives excellent insight into the company's performance and I recommending doing it at least once in different areas in company during the audits. As I said before, the SMS and the companies in Risky Industries are complex entities with complex processes inside and the "Chronology of Events" is giving a picture of how the company is coping with problems in their complex operations. If the company possesses Audit Trail software (or some other type), the method can be checked here. Here, the possibility of changing inputs is very limited, but not impossible. So, the auditor must do it with due diligence!

It is important for this method of auditing to choose a problem (fault, failure, etc.), because the nature of solving problems in Risky Industries gives a clear picture about the company. Daily operations are routine operations and the level of stress is very low there, but dealing with problems gives a unique picture about the true grit of the company.

It is idealistic to assume that bad things will not happen during daily operations and there is nothing wrong with that. But how and why they happen will tell the Regulator how the companies are monitoring safety, and what was done when bad things happened will tell the Regulator how the company deals with safety events!

There is another thing which is important with this method: It needs to be conducted by the Audit Team Leader. Other auditors are busy with particular departments or areas in the company and this method applies to few departments (areas), so the Team Leader is the right person to do it. There is another reason in the favor of Team Leader: He is usually getting information from different auditors about different departments during the audit, so he is the person who can combine this information with the output(s) of "Chronology of Events" method.

9.2.13 Auditing Human Factors (HF)

Noticing any of the Human Factors (HF), will help the auditor to establish the "personality" of the company.* In some Risky Industries (aviation!), the HF are part of the regulatory requirement for establishing the HF program which needs to be monitored and controlled by the Regulator. The HF program is especially requested for pilots and air traffic controllers (ATCos)! But, in the past, it has been noticed that incidents and accidents also happened due to errors which made by different maintenance staff.† So today, there is an ICAO requirement to monitor and control the HF and to provide training for HF to all employees in all subjects which are included in aviation operations.

In general, HF may affect any operation in industry and as such, they are very important in Risky Industries. To audit HF, the auditor must be familiar with all HF and he must take care of all regulatory requirements regarding HF.

Depending on the regulation in the particular industry, if there is a requirement for any aspect of HF, it must be part of the SMS. Some industries are requesting that companies implement a particular program for mitigating HF and some are looking for particular training to make employees aware of HF issues and how to mitigate them.

If there are requirements for HF in regulation, then the company must mention in the SMS how they deal with these requirements. Usually, the lack of a clear understanding of the HF means that companies do not include anything about HF in their SMS. This is a thing which needs to be checked during the Documentation Audit. Whatever is in the SMS regarding HF, must be carefully examined to see is it effective and is it efficient. Later during the On-Site Audit, it must be checked if it is implemented and maintained in daily operations.

In general, the auditor, when he enters any premises in the company must pay attention to:

---

* See Section 5.2 in this book!
† Maintenance staff for aircraft and maintenance staff for ground-based ANS equipment.

(a) The employees (how they behave).

(b) The environment in which the employees work.

(c) The actions they perform during and aside from their duties.

(d) The resources available to the Humans to perform the work.

Things which needs to be checked regarding HF during audits are:

(a) Are the HF included in the SMS documentation (persons, procedures, training, etc.)? (Check the SMS documentation!)

(b) Is there a competent person (doctor, psychiatrist, etc.) who is in charge of HF and/or can provide help to employees if asked? (Check and interview him, if available! Check the records from this person!)

(c) Are the employees in particular positions in accordance with prescribed requirements (education, skills, experience, etc.) for these working positions? (Check the data from the HR Department!)

(d) Are the HF included in Risk Management (HFs identified as hazard, risk calculated for them, investigated during the incidents, etc.)? (Check the list of identified hazards and calculated risks!)

(e) Are the working premises and working environment taking care of HF (enough space, acceptable colors, enough light, good communication capabilities, etc.)? (Check the premises when entering!)

The main point regarding HF auditing is to check the all possible influences inside the company which can affect human performance. Having in mind that there are plenty of such influences in each company, the items to be checked from above (in combination with sampling!) could provide a good guide on how to audit HF.

9.2.14 Auditing Process of Monitoring of SMS

The performance of the implemented SMS must be monitored and controlled by the company. Monitoring is done usually through data-gathering regarding the processes, operations, and events and through voluntary information (Just Culture) or obligatory information system provided by employees.

Each company must have a procedure for monitoring, control, and assessment of the performance of their SMS. The monitoring, control, and assessment of the SMS is part of a proactive approach in safety and may not be neglected.

The auditor should pay attention to the:

(a) Procedures for monitoring and assessment of the SMS performance. (Check the procedures (What, Where, When, by Whom, and How) and their sustainability, are the responsibilities clearly defined, etc.)

(b) Check the implementation of the procedures in reality. (Speak to employees,* check the records, ask for training records, etc.)

(c) Check the daily (monthly) gathering of data regarding the performance of the SMS (look at the records and look for assessment records, etc.).

(d) Check the actions triggered by assessments. (Ask for decisions for change or improvement, ask for the records, etc.)

The auditor must be careful regarding the requirements for monitoring, control, and assessment of the SMS! Noticing many problems in the company does not necessarily mean (in general!) that the SMS of company performs badly. The information on how these "problems" are handled can provide information on if the implemented SMS good or bad! Of course, the good SMS will solve the problems in advance, but the Regulator must give time to the company to establish their SMS.

9.2.15 Auditing Training in the Company

Training in the company is connected to the company's performance. Employees must be properly trained for Operational and System Procedures and also for particular changes in the operations or structure in the company. Training records must exist and in many Risky Industries, there is a regulatory requirement to take proper care of training.

The main point with the company's training is that the auditor must understand that companies are not educational institutions, so they cannot train their staff to achieve an academic degree. If there is a position which requires a person with an engineer background, then this position may not be filled by a technician with the explanation: We trained him. Usually, the training provided inside the companies in conducted by manufacturers of equipment (which is good), by persons from the company with an educational background (which is also good), and/or by the persons who are proven to be good in the operation or process (which may not always be good).

The auditor must understand how the training was executed and he must find proof that it was not just formal† training. He must pay attention to:

(a) Procedure of the training. (Check if it is clear What, Where, When, by Whom, and How it can be done, are the responsibilities clearly defined, etc.)

---

* The auditor must be careful when he speaks to employees! Whatever they say cannot be used as Objective Evidence! What was said by employees can be used only to check documentation.
† Formal (pro forma!) training is training which satisfies the training form only by documentation, but it does not provide particular knowledge or skills to the trainees (employees).

(b) Training documentation.* (Sample and check a few records about the training plan, presentations, training material, etc.)

(c) List of attendees. (Sample a few and check how many of them attended the training, and (especially!) check if there is any employee involved in this operation without the training!)

(d) Training Evaluation Form from employees. (After training, there must be proper assessment or feedback from trainees, look for a few samples.)

The auditor looking at the training should check does the training provides new knowledge, and new skills, and does it improve attitude and experience.

9.2.16 Auditing Handling, Storage, and Shipping

This is a very important area of auditing and it must be considered with the utmost importance! Many of the Risky Industries use materials and processes and produce waste which is dangerous for Humans, assets, and the environment. The methods of handling, storage, and shipping of all these materials and waste makes the difference between good performance and incidents or accidents.

The job of the auditor is to check the documentation (procedures and records) during the Documentation Audit and their implementation and maintenance in reality during the On-Site Audit. Things which must be checked are:

(a) The presence of the procedures which take care of handling, storage, and shipping of dangerous materials and waste. (Check the Manual and other documentation.)

(b) Implementation of the procedures in reality. (Check the training records about the training, speak to employees, ask for the records, ask for PPE, etc.)

(c) Records which are needed to prove that the procedures are maintained in reality. (Check the records on site!)

(d) Monitoring and control over areas where handling, storage, and shipping is executed. (Visit and check these areas (storage!) and especially check the areas where waste is stored and shipped, etc.)

(e) Check the lists of sensors and measurement equipment there, and calibration data for a few of them (if different from the devices explained in Section 9.2.11).

---

* In some Risky Industries, there is a requirement to implement a Training and Compliance Management System in the company, so look for the Manual!

I do not like to look like I am repeating myself, but there is a difference regarding the regulation for alarms in areas of the company which are the subject of manufacturing (covered by Section 9.2.11 in this book) and alarms for areas of handling, storage, and shipping of dangerous materials and waste. The primary difference is that during the manufacturing process, Alarm Systems are measuring a momentary situation which can endanger the production process or Humans, assets, and the environment. In contrast, the Alarm Systems in the storage and shipment areas are measuring cumulative values (which accumulate each day due to keeping dangerous materials inside). These differences are expressed by different tolerances for warnings and alarms. The auditor must have that difference in mind when he is checking all types of alarms!

9.2.17 Auditing Internal Audits

In many Risky Industries, there is a regulatory requirement that company must control itself. As it is written in Section 4.2 (Internal Audits), one of the methods to control themselves is the Internal Audit.

Regarding Internal Audits, the auditor must take care of the following:

(a) Check the training certificates for Internal Auditors in the company. (They must be trained for Internal Audit.)

(b) Check the Internal Audit reports by sampling a few of them and comparing the names of Internal Auditors with the auditors who were trained for Internal Audits. (All of the auditors must have certificates for Internal Auditors!)

(c) Check the Internal Audit reports by sampling a few of them and compare the dates with the annual plan. (They should be in agreement. If there is a difference, ask why this happened. Anyway, the annual plan is a live document and it is not an issue if the dates do not agree. The problem is if the audit is missing (not executed at all!)

(d) Sample one or two Internal Audit reports with findings and check the activities by which the findings were rectified. (It can give an excellent picture about the safety performance of the company!)

(e) Check if the Internal Audit's reports were distributed to Top Management. (There is a requirement in most of the Risky Industries to inform the Top Management regarding the safety performance of the company. It can be a cumulative report distributed at the end of the year, or it can be an Internal Audit Reports distributed to Top Management after each Internal Audit.*)

---

* The ways in which the Internal Auditors submit the Internal Audit Reports differs from company to company. Anyway, whichever way is used, it must be documented in the company's SMS Manual.

(f) Check the response of Top Management to the findings in the Internal Audit Reports. (Ask for approvals regarding the Preventive and Corrective Actions, ask for Follow-Up reports when actions are finished, etc.)

One of the biggest problems with Internal Audits is that these are done mostly as a formality. The auditor must watch out for that.

9.2.18 Auditing Outsourcing and Partner Companies

There is considerable complexity in today's industry. Companies very often use outsourcing or partnership with other companies or organizations. It is understandable that:

(a) At airports (aerodromes, factories, hospitals, etc.), the power supply is provided by companies which take care of the electrical distribution system.
(b) The communication between different aerodromes and air traffic control centers is provided by communication companies.
(c) The storage, shipment, or disposal of nuclear material or nuclear waste from nuclear power plants is very often outsourced to other companies.
(d) Medical waste is usually not destroyed in hospitals, but there are other waste companies which take care of that.
(e) Chemical waste can also be dangerous, so there are outsourced companies which take care of that also.

Of course, there is no reason for the Regulator to audit also these outsourcing or partner companies, but this is an obligation for the audited company! So, the auditor during audits must check:

(a) Is there any annual audit plan for the outsourcing/partner companies? (Ask for the plan!)
(b) Is the plan sustainable and maintained? (Ask for audit records, check for their integrity, etc.)
(c) Check if there were findings and were there follow-up activities. (Sample a few audit records (two to three from each outsourcing/ partner company) and check the dates and outcomes.)

These outsourcing or partner companies are a vital part of the company's operation and they have safety significance in daily activities. Maintaining control over these companies is the responsibility of the audited company. The Regulator must check the records which could prove that every control mechanism is in place.

9.3 Final Report

The Final Report is a document which is produced by the Team Leader and it is distributed approximately two to three weeks after the audit. This length of time is subject to the complexity of the audit. For big companies, the preparation of the Final Report will need more time than for small companies.

It is mentioned in Section 7.5 (How to Present the Findings) that there is an Interim Report which is presented at the Closing Meeting by the Team Leader. This report is called "interim" because it is prepared at the end of audit, in the premises of the company, and it will serve as a basis for the Final Report. As such, the Interim Report is prone to "fast thinking mistakes/errors" and it should be reconsidered later. The Interim Report should be also updated with the comments of the company submitted to the Audit Team during the Closing Meeting. Later, the Audit Team will meet in the Regulator's offices and the Team will reconsider the Interim Report again. On the basis of discussions in this meeting in the Regulator's offices, the Team Leader will produce the Final Report.

There is nothing strange if the Final Report differs from the Interim Report. Actually, it should differ. The main difference is in importance: The Final Report is a legal document and the Interim Report is just the basis for producing this legal document. As a legal document, the Final Report can be used also in the future, in court or for other purposes, depending on the nature of the future Regulator's activities.

Regarding the Interim Report, you should understand that: The remedy for the "love at first sight" is the "second sight!" The "second sight" of the Audit Team into the Interim Report will be done after the audit, in the Regulator's premises, in a relaxed atmosphere, when the pressure of the On-Site Audit is finished. It would provide a more balanced approach to the findings and objective evidence, especially when the comments given by the company are considered.

The most important thing is that the Final Report is not something which can be "democratically achieved." In the case of different opinions about different findings and objective evidence inside the Audit Team, there shall be no voting by the Audit Team. The Team Leader is in charge with the audit, he is the guy who will sign the Final Report, and he has responsibility for the content of the Final Report. So, in case of disagreements inside the Audit Team, the Team Leader is one who will decide what to put in the Final Report. Maybe he will be "stronger" with the findings in some areas and maybe he will be "weaker" in some areas, but he is in charge.

It is good if the Team Leader prepares to present Final Report keeping in mind the previous and future audits. He must think in advance how the report will influence the company and industry at the same time. Producing the Final Report only for this audit and only for this company is a "shortsighted" approach. Assume that the company have introduced some method (tool, system) for Risk Assessment (or for anything else) and the audit showed that it was ineffective or maybe it had caused some incidents. The Regulator must not forbid this method (tool, system) in the company before he analyzes what are the reasons for this ineffectiveness. Forbidding the use of such things in the company cannot be done on an individual basis and must be done through regulation which will affect all industry. Maybe this method (tool, system) in other companies will work. This is a thing which the Regulator cannot know without detailed analysis.

At the beginning of the Final Report, the Team Leader should put a short introduction (Executive Summary is good for such a purpose!) with a general explanation about the finished audit, the Team members, the company, etc. Of course, the updated Audit Schedule needs to be there. "Updated" means: The changes which were made during the audit to the Audit Schedule should be there. This information will give a picture regarding the On-Site Audit activities when someone reads the report later.

As I have already mentioned in Section 7.2 (Findings) and in Axiom 13, the audit is about good things. That is the reason that all good things found in the company during the audit have to be mentioned at the beginning of the Final Report.

I do not propose that the Team Leader must mention all of them: It is simply futile to do that! But the Team Leader has to bring in front of the Final Report the overall impression from the company's performance. It cannot be bad, because even if it is bad, not everything is bad. It is good if some extraordinary things or innovative solutions by the company, are pointed out in this part of Final Report.

The main rule for writing the Final Report is: Always start the Final Report with the good things! Especially in the cases where the company really performs well. It will raise their selfconfidence and it will have good impact on future cooperation. Auditors must not forget Axiom 8: The Regulator and companies are partners in achieving safety in Risky Industries.

Beside the things mentioned above, the Final Report must contain:

(a) All findings and details and comments about them (following particular taxonomy).

(b) All objective evidence connected with each finding.

(c) Detailed consideration of the comments made by the company in the Closing Meeting or during the audit regarding the findings.

(d) Requirement for company to provide detailed explanations to the Regulator regarding Preventive and Corrective Actions (how and when) to fix any particular finding (Safety Cases, if necessary!).

(e) Requirement for company regarding information about the timeline (final date!) from which the information from item (d) shall be provided to the Regulator.

(f) Approximate dates of Follow-Up Audits by the Regulator for each particular finding. (This is subject to the company's fulfillment of requirements to fix the findings.)

(g) Whatever else the Team Leader may find important regarding the company's performance to be stated in the Final Report.

In some of the literature, especially on the internet, you can find that the Final Report has to contain the scope of the audit and the audit methodology* and I do not know what to say about this...

Regarding the scope of the audit, it is useful to have it only for Follow-Up, Special, and Exceptional Audits. When the auditors prepare themselves for audits, they can look in the archive for the past events or audits in the company and they can find information about the performance of the company which will help them to prepare for the upcoming audit. For Approval and Regular Audits, the scope is the overall safety performance of the company. Don't be confused by sampling explained in Section 8.5 (Sampling). It is just a tool which helps to make the audit shorter, but the scope of the audit is still the overall safety performance of the company.

In general, I have no objection if any Regulator decide to produce a Form of how the Final Report should look (for every type of audit) and if the Audit Scope is part of the rules there.

On the internet, you can find also the rule explaining the Audit Methodology in the Final Report...

Regarding the Audit Methodology, it depends on the type of the audit, what type of Risky Industry (aviation, nuclear, chemical, etc.), and what is the size and structure of the company. My humble opinion is that audit methodology needs to be clarified through regulation, but it must be done in a balanced way. Regulation should be done expressing the rules on which methodologies† for audits can be used, but it needs to be general enough to allow the auditor to choose what he finds most appropriate. If the Audit Methodology is determined by regulation, there is no need for it to be explained in the Final Report. Maybe the Team Leader can mention if there is any difference from the regulation or specify which of the methodologies is used (if the regulation determined a few of them).

In general, the shape of the Final Report must be decided by the Regulator and I hope that my opinions stated in this paragraph can help to produce something which fits audit's purpose.

---

* This is common in financial auditing where there are a lot of different methodologies on how to do an audit and I am not sure that it can apply for Safety audits.

† Audit Methodology is something which mostly applies to Financial Audits. I am not aware about Audit Methodology which will apply to Safety Audits in Risky Industries, but if there are any, you may use them.

9.4 What Next?

When the Final Report is ready, it will be printed in a few hard copies. Today there is no issue with producing an electronic copy with signatures for privacy or maybe with a password for restricted use. One (or more) hard copies can be sent to the company and the Team Leader will wait for the response of the company.

The response of the company will be given in the form of an official document where the company will:

(a) Comment on the findings and/or objective evidence.
(b) Provide a schedule for each finding showing how it will be rectified. (Simple explanation, or submission of a Safety Case if finding is considerable!)
(c) Their agreements (or objections) regarding the "deadlines" for fixing non-compliances submitted by Regulator.
(d) Anything else that they found important regarding their performance in the audit.

Commenting on findings and objective evidence by the company is a normal thing. They may not agree, but they must fix the findings from Level 1 or Level 2. Providing a schedule of how they will fix the findings will need time, so the Regulator must give them a reasonable timespan for this proposal. Having in mind that these findings were presented in the Closing Meeting which happened two or three weeks before the Final Report, the company is aware what is going on. So, the under the term "reasonable timespan," I mean a time not more than two weeks from the distribution of the Final Report to the company's response. There is a situation in which they could ask for more, but the company, in such a case, must submit a reasonable explanation. The Regulator (Team Leader) will consider this request for prolongation of these two weeks, and there is nothing wrong with him allowing them more time.

9.5 The Regulator and the Outcomes from the Audits

The audit is finished for the company and the next one will be in 6 to 12 months; but it does not mean that the Regulator's job is finished. The audit provides excellent insight into the company's performance in areas of safety, but there are many companies and many audits for each of them. The Regulator finishing the audit in one company gets a picture from only this company and now they need to be summarized to see the overall situation and perspective of all industry in the State.

Each Regulator must understand that the Final Reports from different audits of different companies must be analyzed with the intention to see how the overall industry in the State performs. This summarized analysis of all the findings can provide extremely useful information about systematic errors in regulation or in industry.

It is necessary for the Regulator to statistically assess the data regarding the findings. This analysis (which is internal and local by volume!) should be compared with the analysis on the global level (other States or regions!). The statistics could provide good insight into what are the trends in industry in the State, are they in line with global development, and is there something which can be done regarding the regulation being changed or improved. The weak points in regulation or in oversight activities of the Regulator, are evident if the same findings are found in few companies. In addition, the Regulator must analyze and correlate results from this year and from previous years and the results must be used to produce some conclusions or actions. If the audits are to provide a picture of the company's performance, the later analysis of the all Final Reports, will give a picture of what is going on with the regulation in this area, in this State or in this Risky Industry.

Some of the results of this analysis will provide answers to these questions:

(a) Is the regulation satisfactory?

(b) Is there a wrong understanding of regulation by the companies?

(c) What is the trend in industry in the State?

(d) Is the trend in line with global development?

(e) What will be the further actions of the Regulator (on a local or global level)?

Having some discrepancies between expectations and the real situation with the safety performance of the companies is a problem for Regulator which may not be neglected. And have in mind that problems are like children: They grow up very fast and one day they will become bigger than you and they will slap you! And it will hurt! So, the problems need to be solved as soon as they are registered. This is one of the aspects of good safety performance for the companies and for the Regulators.

If the Regulator does not analyze the Final Reports, it means that he does not have an understanding of how good the regulation is and what is going on in the industry.

10

"Challenges" for the Auditor during Audits

10.1 Introduction

An auditor during his job has a lot of problems. One of my colleagues use to say that there are no problems, only *challenges* in the human life. And I like this attitude: Problems ("challenges") exist in human lives only to be solved! The way we solve the problems is what we are! But with regard to audits, let me paraphrase: The way the companies solve the problems is what the companies are!

Some of the possible "challenges" regarding a Documentation Audit were mentioned previously, and here I will stick to the "challenges" which can happen during On-Site Audit.

Having in mind that an On-Site Audit is conducted on real time, these "challenges" must be also solved in real time. It does not mean that the auditor must immediately offer a solution, but it means that he has to be prepared for them. I will mention here a few of these "challenges" which I had the chance to notice.

10.2 Time Gap

This is a common problem in all audits. Very often, the companies are trying to put the auditor in the so-called "Time Gap." It means that they will try to prolong audits in the areas where they feel confident, and it will give less time for the auditor in the areas where they feel the auditor can find something wrong. The Time Gap will cause the auditor to have less time for the audit in such areas, so in the rush to stick to the schedule, the auditor would neglect something which could result as non-conformance. In other words, due to lack of time, the audit in these areas will be just a formality.

How can this "challenge" be manifested?

Usually, there are a few such "tricks" used by the companies:

(a) The company's person who is dedicated to the auditor is offering to show the auditor something which is not part of the Audit Schedule.

(b) The same person is trying to provoke conversation with the auditor which has nothing to do with the audit.

(c) The same person will provide free lunch for auditor and he will try to make it to last very long. Usually you go to a restaurant close to the company (or on the company premises) and you are waiting too long for the food.

(d) The person who needs to show some documents (records, lists, measurements, etc.) is not in the office (lunch, toilet, or he is sick).

(e) There is no transport or it came too late to visit a particular department which is far away from the Main Building.

I must say here: It is maybe not realistic for the companies to fear that something is wrong with their Safety Management System (SMS). "Not realistic" means that they feel unconfident in some areas, not because they did something wrong, but due to some other reasons. Maybe in these areas, the regulation is not clear and their understanding of the regulation differs from that offered by Regulator. Maybe they had some observations in the past audits, and they did nothing to rectify them (nevertheless it is not mandatory to deal with observations, which are Level 3 Findings). Maybe there is something which is part of their research activities and they do not like auditor to see that.

As I said, if such a thing happens, the auditor must not jump to the conclusion that something is wrong. There could be many reasons for these things to happen. But maybe really, there is something which is non-compliant there...

Anyway, the auditor must be alerted that the audit in some areas is taking too much time (or the audit is close to the Time Gap!) and he must understand if it is intentionally or nonintentionally done.

Whenever such a thing happens, the auditor must stick to the Audit Schedule! But how can he handle this challenge?

To prevent such cases, it is important for the Team Leader to think in advance! Thinking in advance means that while creating the Audit Schedule (before the On-Site Audit!), the Team Leader must provide excess time for each auditing areas. This could be especially important if the Audit Team is going for the first time to audit the company. In such a situation, the Audit Team is not aware of the company and they are not familiar with all the structure and organization inside the company. For example, when auditing Air Navigation Service Provider (ANSP) in aviation, auditor will visit at least one or a few sites where navigation or radar equipment is installed that are not in the aerodrome area. Usually they are outside, maybe on the top of a mountain and the auditor needs an SUV (Sport Utility Vehicle) to reach it. In simple words, the Team Leader could not know how many times they will need to reach (or audit) particular areas.

In general, it is wise if for any calculated time, the Team Leader adds an additional 10% of the time as a "back-up." He must have in mind that some of the audits in some areas will finish faster, so allowing 10% more time for each area, on average, could provide good margins for sticking to the Audit Schedule.

Second thing is, there is no need for auditors to worry how the company will react to the audit's requirements: They need to fulfill all reasonable requirements of the Audit Team. As explained before (section 8.4.2 – Preparing On-Site Audit), the Team Leader will send a letter to the company informing them about the date of On-Site Audit and providing to them the proposal of Audit Schedule. So, the company is informed in advance how the audit will be conducted before the audit starts. This is time when the company may comment on the Audit Schedule and they may suggest some changes in the Audit Schedule. If they do not do it, it is assumed that they agree with Audit Schedule.

If they submit some comments and suggestions regarding the Audit Schedule, these need to be considered with due attention by the Audit Team. Whatever is decided, it must be communicated to the company with due politeness and with explanations of the reasons why it was refused (if refused!).

So, each requirement regarding auditing some areas in the company from the auditors is known to the company in advance and they must provide access in these areas in advance. As the Audit Team needs to prepare for audit, so the company must do it!

The Team Leader and the auditors (members of the Audit Team) must discuss with the company (again!) the Audit Schedule at the Opening Meeting. Simply, in the meantime, something could change and it must be taken into consideration. But the Team Leader must clearly explain that when the Audit Schedule is established (confirmed!) at Opening Meeting, the Audit Team and the company must do their best to stick to that!

Anyway, if during the audit, the auditor notices some attempts at "creating a Time Gap" by the Counterpart, he must (politely!) explain to the Counterpart that they must stick to the Audit Schedule and whatever happens with the audit, it is company's fault. If the lunch (or some refreshment) is offered to the auditor, he must explain that in 30 minutes (or 10 minutes), the audit must continue. If the danger of the Time Gap is still present, the auditor must leave the lunch (refreshment), politely explaining that there is a job to be done. Under no circumstances may the auditor endanger activities in the Audit Schedule.

But what to do if, in any case, the audit cannot finish properly by the Audit Schedule?

These things happen and they need to be considered at the meeting of Audit Team. If there is serious doubt about the company performance, the Audit Team may decide to provide a Special or Exceptional Audit to the company in the next few weeks. Anyway, it must be stated in the Final Report that due to some reasons (there is need to prepare list of the reasons!), the audit did not cover particular areas. Of course, it will not be mentioned in the Final Report that there will be Special or Exceptional Audit in the future to cover these areas!

This is something which will be decided by the Regulator in the case of serious doubts regarding the company's performance.

10.3 Not Following the Audit Schedule

It is not only the Time Gap that could produce problems with sticking to Audit Schedule. Sometimes, the planned time is not long enough to do the audit properly. Such a thing happens sometimes, but it is common for audits which are done for first time in any company.

When the Regulator does not have experience with the company, the Team Leader can produce an Audit Schedule which will be overestimated or underestimated. An overestimated Audit Schedule means that in some areas, the auditor needs less time than planned for proper auditing. Underestimated means that the auditor, in some areas, needs more time than planned for proper auditing. Overestimated, if it happens, is not an issue. Usually, there is a lack of time for an audit, so the auditor can use this time for something else. But an underestimated Audit Schedule can be a problem, especially if it is extensively underestimated (there is lack of time in a few areas of auditing).

That is the reason that my advice from previous paragraph for a 10% increase of the necessary time for auditing in some areas should be accepted literally. Whatever time is dedicated to a particular area in the Audit Schedule, the auditor must manage it properly.

I do believe that this looks for some too bureaucratic, but believe me: The problems which arise with this "bureaucracy" are far less than the problems created if the Audit Team does not stick to the Audit Schedule!

10.4 Company Does Not Allow Access to Some Premises

There are situations where the audited company will not allow access to the auditors to some premises. This could happen in some companies and they affect the areas of research and development where the new technologies or products are tested.

Whatever the area is, it must not happen!

The Regulator and its auditors must have access to everything. That is the reason that, if a Security Clearance is requested, they must have it. It shall be part of the regulation where the "rules of engagement" of the auditors and company during the audits must be precisely regulated. Anyway, the

Regulator must assure each company that he is a reliable partner, and there is no chance that any company secret will be endangered. In some companies, the auditors can be requested to sign a Non-Disclosure Agreement and there is nothing wrong with that: The auditors should sign it!

But the main point is that the auditor must find the real reason why the access is not granted. Some of the companies, especially these with "personality" Class 4 and 5, can use this as an excuse in the areas where they do not comply with the regulatory requirements. So, if there is something in the Audit Schedule to be visited, the company must provide access. If the access is not granted, the information about that should be revealed before the audit.

If such a thing happens, the Lead Auditor must warn the company's counterpart that the access must be granted. If the access is not granted, then the Audit Team must stop the audit, and the Regulator in such cases must stop operation of the company. It is maybe too strong a solution, but having in mind that the Regulator has provided all assurances to the company and the Audit Schedule was given in advance to the company, the access must be granted. Obviously, there is something which could raise the alarm with the Regulator and stopping company's operations is only solution. It must be done by force, in the presence of police, if necessary.

10.5 Company Is Lying

During my education in Management System areas, in one of the trainings which I attended, the instructor mentioned the situation of what to do if the auditor finds that the audited company (or some of the employees there!) are lying to the him during the audit. It was training for Lead Auditors (Team Leaders) and he stated that if this happens, the On-Site Audit must be stopped and auditors must leave the company. Simply, in his opinion, an audit is a matter of confidence between two parties (auditor and auditee), where both of them should earn benefit. So, if there is lying between the partners, we cannot speak about partnership. Such a thing could clearly happen with the companies of Class 4 and 5, maybe even Class 3 from the Table 5.1, but in general, small lies (especially non-intentional ones) could happen with each class of company.

If the Audit Team decide to leave the company, the Team Leader will submit his report (why the Audit Team left!) to the Regulator, and in such a situation it is normal that some (or all!) of the company Post-holders (CEO, General Manager, and/or Safety Manager) will be deprived of these titles by the Regulator.

I am a little bit reserved in the matter of such a "revolutionary step" as to leave the On-Site Audit, but honestly, I do believe that it is a good solution.

My "reservation" comes from my attitude that I, in general, prefer "evolution" instead of "revolution" and leaving the audit is "revolution" for sure!

But speaking about auditing SMS and finding that the company is lying (cheating!) to you, in an industry where the consequences from wrong performance can be extremely terrible, this made me to accept the step of stopping with the On-Site Audit and leaving the company immediately.

What I would like to recommend here means that auditor must be careful with those things and the decision to stop the audit must be decided at the Team Meeting where the Team Leader and other auditors will reconsider all aspects of the dishonesty. The recommendation to do it as a Team has nothing to do with the possibility of forgiving the company. Auditors shall not put themselves in a situation to forgive somebody for something. In each such situation, (again!) the decision should be made on the basis of objective evidence. The auditors must understand that in such a case, there is a high possibility that the company will try to sue the Regulator and the reason for stopping the audit (and not granting permission to operate) will need to be explained in court.

Anyway, I am encouraging such a step (to stop the On-Site Audit and leave the company premises), but each Regulator must in advance consider such a step in general, before any audit activity. It is good if companies are informed in advance what will happen if lying occurs, and it is good if this is mentioned in the regulation, maybe not by law, but by information submitted as a Safety Circular* to each of them.

10.6 Auditor Found Something Which Is Not in Target CLs

During On-Site Audits, the auditors are "wandering around" inside the company's premises and they have the chance to see different situations and experience different events. It could happen that during the On-Site Audits, some of the auditors notice something which is not in accordance with the regulation, but is not mentioned in the Target CL.† So, the question here is: What should the auditor do in such a situation?

If he tries to investigate this notice, it will deviate from the Audit Schedule and it will create a Time Gap for sure. But if he neglects it, it is wrong: It may have terrible consequences in the future.

---

* Some Risky Industries (aviation!) use an official document named a Circular as a tool of disseminating information about something which is not in the regulation, but relates to safety, technical, administrative, or legislative matters. The companies which receive the Circular must take into consideration the content of the document.

† If it is not mentioned in the Target CL for this particular audit, it means that it is not subject of the audit.

The best way is to take a note regarding the situation (thing, event, etc.) where it was noticed, what time, what was the issue, and anything else which the auditor finds it important, and to present this to the Audit Team later during the Team Meeting. Depending on the nature of the situation (thing, event, etc.), the Audit Team will discuss it and the Team Leader may decide how to proceed.

If the situation (thing, event, etc.) does not need an immediate response, its investigation may be postponed. In general, the Audit Team must dedicate itself to the Audit Schedule and these situations (things, events, etc.) could produce deviation, which is not good at all. The noticed situation (thing, event, etc.) can be subject later to a new audit (Special Audit) or it may even be postponed for the next Regular Audit. A balanced approach is also important in such a situation and maybe the company can be informed about this notice. In the most of the cases, it will help the company to solve the issue itself. Anyway, the next Regular Audit must check again the follow-up of the situation (thing, event, etc.).

But if there is immediate Risk regarding Humans, assets, and the environment, the audit must stop and the particular Corrective Action(s) must be implemented immediately! The Audit Team must be present to investigate why this happened and what was the response of the company to this situation (thing, event, etc.). The follow-up activities from the Regulator and the company could be established after the situation (thing, event, etc.) are again brought to control. The Team Leader (in consultation with the Audit Team) will decide should the audit continue or if it will be postponed for a particular period of time.

10.7 There Is Not Enough Staff in Company to Maintain Safe Operations

As I mentioned in Section 3.2 (How to pass the regulation?), the satisfying of the regulation is a burden for the companies! The companies must employ people and/or must use other resources. It is always costly and it is time-consuming for them.

The one (also) critical thing with the audits, which needs to be considered by the Regulator is: Is there enough staff to maintain a safe operation. And this is not an easy job!

As I have said before (in Section 2.11, Establishing a Management System), companies are in charge of employing Humans (staff) to conduct their operation. The profile of the staff will depend on the nature of the operations, but the number of staff usually depends on the assumptions of the company's managers. They know that more employees will mean more costs and when costs rise, profit goes down. So, having not enough staff will mean that stress as a Human Factor is present in the company and it will increase the probability of wrongdoing. This is an issue and the Regulator must take note of these situations.

There is one bad example regarding understaffing in companies in Risky Industries which happened in aviation. On the night of 1 July 2002, two aircraft collided over the German town of Überlingen. A few months before the accident, the Swiss ANSP (Air Navigation Service Provider) Skyguide, which was in charge of providing ATC services that night, was privatized. With the intention to save money, Skyguide decided to change the organization of the company. So, in the scope of changes, they decided that, during night shifts and not so big air traffic in the region over Überlingen, the separation of aircraft can be provided by only one ATCo (Air Traffic Controller). But, one of the findings of the investigation* was that having a second ATCo in position that night could have considerably affected the situation.

So, companies in each industry are trying to save money and increase profit (efficiency!), but understaffing in other industries may affect only the company's incomings and maybe its position in the market. In Risky Industries, it can cause incidents and accidents.

The real problem is that the Regulator cannot and may not establish the number of employees which can provide safe operations for each company. Even the Regulator must not state that there is not enough staff for a safe operation if there is no objective evidence! The point is that even I cannot recommend anything which will help the Regulator to handle this problem.

Anyway, there is a solution, but it will depend on the type of Risky Industry and from the Regulator's capability to "read between the lines."† There is no direct way to establish that there is a lack of staff in a particular company, but it can be achieved in an indirect way. The company's staff is connected with activities inside the company, so checking the time performance of a particular operation could indicate the lack of the staff. For example:

(a) Checking of MTTR‡ of a few corrective maintenance actions by the auditor will show how the company is fixing faults in their equipment. (This is connected with the failures of their operations!) Having in mind that the MTTR should be not more than 30 minutes, exceeding this time could be an indicator of understaffing. But this situation needs to be investigated before jumping to a conclusion. Maybe this is a failure of maintenance training.

---

* To be honest, there were a lot of factors which contributed to the accident. In my humble opinion, the main reason for this accident was the incorrect understanding of how to use TCAS (Traffic Collision Avoidance System) by one of the crew. Anyway, everybody agreed that having two ATCos that night could have prevented the accident.

† The meaning of the expression "read between the lines" is that the auditors must look for the hidden meanings of activities inside the company.

‡ MTTR stands for Mean Time To Repair. This is time needed by the maintenance staff to fix an unexpected problem (fault!) with equipment.

(b) Checking any type of training performance by the auditor could also show a lack of staff. The trainees which will attend the trainings are employees who are doing daily operations in the company. So, bringing them on training means they will need to be replaced in daily operations with someone else. If there is enough staff, the trainings will be regular and there is no need to repeat the trainings more than once. It means that if there is enough staff, the employees will be divided into two group with the same number of trainees and each group will be provided by one training session. If there is not enough staff, the training groups will be not more than two and the same trainings will be repeated few times.

(c) In Risky Industries where the operations are executed in 24/7 manner, there is a need for shifts based on the shift durations. This can be regulated, but anyway I can just give few considerations here. The point is that it also depends on the situation in the company. The Regulator can handle these situations using the example here: If the shift working time is twelve hours, three shifts are enough, but if the shift working time is eight hours, then four shifts are necessary. In the case of twelve-hour shifts, there is need for at least five employees (two for shifts, one on rest, and two as back-up in the case of illness, leave, or some other absence). In the case of eight-hour shifts, there is need for at least six employees (three for shifts, one for rest, and two as back-up* in the case of illness, leave, or some other absence). I must repeat here again: This is something which Regulator can regulate and he will consider all "pros" and "cons" during this consideration.

(d) Due to economic reasons, in many companies from Risky Industries, the monitoring and control is done by the same staff which is dealing with maintenance. In such cases, the auditors can check a few corrective maintenance records and see how many persons were involved in the maintenance (during these faults) and how many persons were left in Monitoring and Control Room in the same time. The company must document the number of staffs needed for monitoring and control and for maintenance, so it can be checked by the auditors.

(e) Checking the overtime hours can also be an indicator of understaffing. Having a lot of overtime hours, is an indicator that there is not enough staff to deal with the problems. If there is enough staff, there is no need for employees to stay overtime: The next shift will simply undertake the problem-solving.

---

* There are companies which use only one employee as back-up, but I strongly object this solution. If someone from the shift is on leave, he will be replaced by the guy who is back-up. But what will happen if (at the same time) someone gets ill? Having two guys as back-up in each shift is the wise solution!

So, in general, there are indicators regarding understaffing in the company, but the problem is to connect these indicators with objective evidence. Anyway, the Regulator may mention understaffing in the Final Report and it can be done as finding Level 3 (Observation). If this is mentioned in any Final Report of Regular or Approval Audit, and if something bad happens in the future which is connected with lack of staff, the company cannot deny that they were warned.

The Regulator should be careful in dealing with understaffing in the companies.

11

Profile of the Auditor

11.1 Introduction

It is not necessary for all auditors to be specialists in the particular area of auditing. This is good, but it is not a must! For Documentation Audits, auditors do not have contact with the company's employees. They are judging the company's Safety Management System (SMS) only by that which is written in the documentation. But for On-Site Audits, they need more psychological training and they judge not only the system, but also the employees who are taking care for the system. Maybe it looks strange, but you can judge the capabilities of the sports team depending from the players and manager appointed. Similar things happen in every aspect of our lives: We judge things by how they show up and their behavior. So, why not to use it to establish the "personality" of the company in the audits also?

But be careful! Auditing is not about the capabilities of the employees! It is about what is achieved using these capabilities!

And the same thing happens even in sports: Very often, total outsiders win the championship. So, do not forget: However, the knowledge about the "personality" of company and knowledge about employees can help you, do not use these things to judge the SMS! Do not forget that these are just things which could help auditors to build a strategy and tactics for audits. This knowledge does not provide information about the compliance of the Management System audited, but it provides knowledge on how to behave during the audits and what to ask!

Axiom 22: Establishing the "Personality" of the company by the auditor is needed to provide information how the auditor will behave and what questions and when to ask them during On-Site Audits! The "Personality" of the company may not affect the findings!

11.2 What Is Important for the Auditor?

The auditor should not be a person who is an expert in the particular area. There is good reason for that! If we can employ an expert to be auditor, this is OK, but we need more expertise in the auditing field. This is a good place to repeat the statement from Section 2.1 in this book:

> "The Auditor does not need to be a chicken to lay an egg, but he must have the capabilities and knowledge to determine which egg is good and which one is bad!"

If there is a need to have deep knowledge about the issue (and it could happen!), then the auditor may (and should!) engage a specialist! There is nothing wrong with that. It is not necessary that the auditor must know how the Risk Assessment is made, but he will have knowledge in advance about the expected values for particular risks. And if these values are not OK, then maybe the specialist in Risk Assessment methodologies can check if everything is OK with the method used for Risk Assessment in the company. It is not an easy job, because it involves statistics, probability, Boolean algebra, etc. So, the specialist is the right person for that!

Let's see why it is more important for the auditor to know the people and how to behave with them.

I will present a good example from security in Israeli airports. A terrorist attack or aircraft hijacking never happened in Israel and there is very good reason for that.

They apply all the measures known for airport security, such as CCTV cameras and X-rays for checking luggage and humans, and they have one more thing: They have established a few additional checkpoints (up to five) before the passengers enter the airport. There are persons on these checkpoints who are trained in human behavior. They can easily register the stress or the anxiety which can be present with humans when they try to do something wrong. They will not bother you; they will just ask simple questions which are slightly changed at each checkpoint. If there is any anxiety in your answers, they can notice it and they will raise the alarm. Of course, you will not be arrested, but you will be subject to additional checks. And if there is something behind it, they will find it, just due to your wrong intentions which are reflected in your behavior! The passengers with wrong intentions would not be able to deal with the pressure of these investigations on the checkpoints and they will betray themselves. And the Israeli guys who do all these checks are really good!

This is something which is important for auditor: If he is good with humans, he can notice any intention of the company's employees to hide something or to lie to him. So, there is no need for expertise in the area of technology (systems, equipment, theory behind them, etc.), but there is a need to know how people behave when there is something wrong.

Do not forget: Management is about managing people, not about managing machines (systems, equipment, etc.). "Managing" equipment is called engineering!

Another thing is that the auditor must know is that "more flies can be caught with honey than with vinegar." He must always be polite, but behind this politeness, there should be a hidden intention to understand (in a subtle way!) if there is something good or something wrong.

How it can be achieved?

There is one way to do it:

If you analyze your conversation with humans, you can notice that this is usually predictable: We usually ask the same questions and we usually get the same answers. When we meet someone, we usually ask: Hi, how are you (how is it going)? And we get always the same answer: Hi, I am OK (it is OK)! Even our counterpart in the conversation do not feel OK, by automatism, he will provide this answer: I am OK (it is OK)! This is a habit which is in our minds and humans use these habits to behave routinely in every situation. In general, humans like routine because it makes them feel comfortable in every situation!

But if we ask a different unconventional question (which cannot be answered as routine), it will not be in accordance with the expectation of our counterpart and he will be confused (especially in the cases when something is wrong!). He will not know how to answer and his respond will be guided by his instincts. And if something is wrong, his instincts are guided by stress and anxiety. Trained person can notice that and he will know that something is not OK.

This is the method used by Israeli security! If the police stop you, your expectations are that they will ask you something which could be wrong. But if they ask simple, polite questions, you will try to adapt your answers and make them friends. And in the ongoing conversation, if something is wrong, you will be catch in the lie. After that, everything is easy.

Now, when something is noticed to be wrong, the auditor must find (again) politely a way how to find what is wrong. He should try to ask different questions in the direction of what was established by the first few unconventional questions. There is no rule on what to ask, but these questions will depend strongly on the answers of the auditee. If the auditee is "pressed" in the right direction, he will soon tell what is the problem.

And this is something which make the difference between good and bad auditor: The good auditor will find a civilized way to find what is going right and what is going wrong!

11.3 Yearly Assessment of Auditors

In the industry, auditors, after passing the exam, they receive a certificate. The validity of the auditor's certificate is usually three years. During these three years, the auditor must conduct particular a number of audits in the area and the scope which is determined by the auditor's certificate. If this number of audits is not maintained, the auditor will lose his certificate. There is nothing wrong with that and I support it. But in some Risky Industries, there are international or national organizations which require assessment of auditors on yearly basis. And I am strongly against this practice!

I assume that the reason for such a requirement is that they would like to strengthen the position of auditor, but it can only increase the stress of auditor.

There is similar situation in aviation. Such checks are provided through refreshment training for pilots and Air Traffic Controllers (ATCos) in aviation. Usually they are associated with some assessments. But these refreshments and assessments are not dealing with daily operations which are regularly conducted by these professions. Simply: If I am doing my job every day successfully, why would I need such a refreshment training?

The answer of this question is recognized by "good people" in aviation, so the pilots and ATCos are subjects on refreshment training which is conducted in a simulator. There, during refreshment training, they refresh only *emergency* procedures for situations which do not happen very often and can be easily forgotten. So, in the simulator, the pilots are refreshing procedures on how to get an aircraft out of stalling, how to fly and/or land an aircraft if one engine is in fire, how to recover an aircraft from bad weather, etc. ATCos are trained how to free the airspace from neighboring aircrafts if some aircraft is in distress, how to change the structure of airspace if there is an emergency situation, how to handle air traffic which is bigger than capacity of airspace, etc. What is important is each of these refreshment trainings are finishing with assessment of the trainees. And it shall be an assessment for such situations!

But, to have assessment for auditor for something which he is using every day is useless and it cannot provide any benefit. As I said, it can only complicate the auditor's job and it can just produce additional stress for the auditor. Keep in mind that the auditor's job is already stressful enough!

I am not against refreshment training for auditors, but only in situations when there is an event when the auditor has made a mistake in something which is important for regulatory affairs. Anyway, I have serious doubts even such situation, because it can be just a nonintentional mistake which could happen to everybody. If it is done intentionally, then the auditor shall be sanctioned (even fired!), but the assessment and refreshment training will not help in such situations.

So, my humble recommendation to every Regulator: Do not bother auditors with regular checks (assessments!)! If you would like to help, provide them training about Human behavior!

The most important thing about auditors is that: "Practice makes them perfect!"

11.4 Improvements of Auditors

I do not like periodical auditor's assessments, but I strongly recommend actions which can improve an auditor's competence by increasing their knowledge and their skills.

This is something which has to be implemented regarding auditors in the Regulators. It is good if they have chance to attend some workshops, conferences, symposiums, and whatever else is available in their area of working. The Regulator actually should encourage active participation of his employees in such events. By active participation I mean, submitting their own papers to conferences and symposiums or their own articles to magazines and journals from particular areas of interests. Purchasing these magazines and journals, and their availability in the Regulator's premises, can provide the opportunity for spreading the view and attitude of the auditors and gathering information on what is going on in the particular Risky Industry.

Attending the training organized by international organizations in Risky Industry is also a good opportunity to catch up with the information about new developments, new regulations, new technologies, and/or new methods and methodologies. A good Regulator will take care over upgrading the capabilities (knowledge, skills, attitudes, etc.) of his auditors and there must be planned money in the budget for such activities.

Someone maybe will say: It is too expensive!

On such questions, my response is always same: If you think that the knowledge is expensive, please check the price of the ignorance! The mistakes done by the Regulator's employees do not affect one company: They affect all industry in particular State!

11.5 Things for Which a Good Auditor Must Take Care

In many countries, the Risky Industries are part of State Administration and they are connected also with the military. In some nuclear power plants, there are research activities by the Institutes or Military units, in aviation, the same runway is used for civil and military aircraft, etc. In all these situations, maybe the State (the Regulator) will need to provide Security Clearance for the auditors. There is nothing wrong with this activity, because it is required. So, the entity which will submit the documents for the Security Clearance shall be the Regulator. It is clear that the auditor without Security Clearance may not be permitted to do an audit in the premises (companies!) where the Security Clearance is required.

The auditor must be properly dressed. He must not be casually dressed. His clothes are part of the picture of integrity which he is presenting about himself and about the Regulator. There are regulators who have uniforms for their officers and if this is not the case, the auditor must be dressed in a suit. The suit must not be expensive and there are two reasons for that:

> The first reason is that it can provoke wealth inequality between the auditors and employees in the company. Do not forget that during On-Site Audits, the auditor is visiting different departments or units in the company, and although the employees in the Risky Industries are well-paid, you may not show them that you have more money than them.

The second reason is that the working conditions in some of these departments are not appropriate for expensive suits, as the suit can be simply damaged (stains, paint, cuts, dirtiness, etc.). It is good for auditor to have a working topcoat with him. Usually in such places, the company will provide protective helmets and glasses, but bringing his own protective boots and topcoats is wise solution.

The auditor may not be an expert in a particular area, but if he is, it is good because it will increase his integrity and respect from the company's employees during audits. Anyway, it is not wise during the audit to speak about expertise in particular processes. The reason is simple: However good your knowledge is, the employees there are working every day in this area and they are masters for sure of their equipment and their processes! If the auditor decides to speak, he must try to ask the questions and not to provide his comments with intentions to show how good his knowledge is. Sometimes it is wise for auditor to write a note about what has been said to him and later, he may check by himself or ask the specialist about that. The auditor must just be sure that everything what is said by company's employees is correct.

There is a belief that it is good if the auditor does not have excellent knowledge about areas of auditing, but I oppose this belief. Some people think that the auditor may not interfere in the company's processes, but being ignorant is not a way to achieve it. If the auditor is not knowledgeable, he may ask simple questions, which is good because it will give him basic information about the things which he is looking. Anyway, to get familiar with the company's processes, the auditor needs to prepare for the audit in advance.

I do believe that it is better if the auditor possesses excellent knowledge in the areas of audits, but he must not show it always. Someone will say that this is cheating, but it does not mean that the auditor is lying. He just may not allow himself to be involved in such "scientific" conversations with the company's employees. In such situations, the company employees will pay more attention what they will say in front of the auditor, simply having in mind that they do not know auditor's level of understanding about their processes. From another side, if the employees feel that the auditor is ignorant, they may lose respect for the auditor, which again is not good.

Both approaches have good and bad sides and it is good if auditor is capable of maintaining both of them.

But there is something which each auditor may not forget: auditing is a complex process and especially the On-Site Audit is extremely demanding in regard to auditor's psychological abilities. So, having in mind that I already spoke about Human Factors (HF)*, he must take care for the HF also in regard to him and his performance during the audits. As the HF may affect everybody, they can affect the auditors also.

So, the auditor must come rested and fresh to the company's premises on the day of the audit. He must have had a good sleep the previous night, he must not be tired, and he must not drink alcohol night before or undertake any medication which could make him dazed. Having a breakfast with carbohydrates (cereals, honey, etc.) is a good choice, because the energy for the brain is really needed during the audits. Coffee or tea is the choice of the person. The day before the audit, the auditor must do whatever he feels will keep him relaxed, but focused and clear-minded during the audit.

Maintaining a good mood and good spirits is also very important during the audits.†

11.6 Qualities of Auditor

There are particular qualities which need to be possessed by the auditor, so he can use them to do his job successfully. Not all of them should be connected to his personality, because some of them can be learned by experience.

When I spoke about these qualities during my theoretical training for auditors in Indian Aviation Academy in New Delhi (2011), one of the trainees said to me:

> "It is clear why we need an Audit Team! One person could not have all these qualities."

It made me laugh (and other trainees laugh also!), but the auditor must strive to learn and implement all of these qualities and most of them must be used whatever the situation is!

The audit is a complex and demanding activity (mentally and physically), but the qualities which needs to be possessed by an auditor can help to deliver a good quality audit. By "good quality audit," I mean an audit which is realistic, not biased, and not compromised!

I know that there are a few Regulators and international organizations for audits which have established a particular document named "Code of Ethics" which must be signed by each auditor and I like it. The auditors who

---

* See Section 2.17 (Human Factors (HF)) in this book!
† See the end of Section 11.6.6 (Communication) and Axiom 25 in this book!

breach this Code of Ethics can be fired (subject to the level of breach!) or can lose the membership in audit organization. Although I support any Code of Ethics, I have noticed that the membership in these auditor's organizations does not mean that the members are good auditors!

Let's continue and explain the required qualities for auditor.

11.6.1 Training in Audits

It is not important for an auditor to be an expert in the area which he is auditing, but he must be trained in audits. And I hope this book can provide enough information for such a training! Audit training will provide the auditor with the skills to get answers to the questions which are of interest to see if the compliance of the company with the regulation is established or not. This is something which cannot be learned by the book and it will be improved by each training and each audit.

It is simple as driving a car: There are many books on how to drive a car (press the clutch, put it in gear, and press the throttle…), but let me ask: Do you know someone who has read these books and then just drives a car? Of course, reading books will not make you successful in driving a car! Everybody must train by driving the car and, in approximately 15 days, he can submit his request to pass the exam with no guarantee for success.

Auditing is a different area than driving a car, but the level of skills must be achieved on the same way: By on-the-job training! It is important to understand that there are rules which must be used during audits and after that, it is all matter of training and experience. Theory (this book!) will provide the rules and explanations about them, but the On-Site Audit (On-the-Job Training!) will provide realistic and sustainable training for an auditor.

Similar to driving a car, in auditing "practice makes perfect!"

Axiom 23: A person without proper audit training may not do auditing!

Having theoretical training will make the auditor good, but reaching a particular level of experience by practice will make him excellent!

11.6.2 Time Management

An auditor must be good in "time management." As I have mentioned in the previous chapter (Chapter 10), there is a possibility when (due to one or few reasons), the auditors cannot maintain the Audit Schedule. This is not good due to one very important reason:

Not maintaining Audit Schedule (usually being late!) can produce a situation when the auditors are struggling to "stay on track," so they try to "speed-up" the audit in some areas. And this is wrong!

The audit should be a comprehensive tool for Oversight. Trying to "speed-up" the audit can produce only bad Final Report. A situation when auditors put in an effort to finish the audit in some areas faster than planned will increase stress of the auditors. Stress (as one of the Human Factors) will make the auditors prone to mistakes, they may neglect their instincts, and they cannot notice some other irregularities or uncertainties with the company's SMS.

Even a good plan for Audit Schedule can be endangered by the bad time management, so the auditor must strive to have control over the process of auditing. It is good to adjust the timer on your mobile phone (the alarm!) which will warn you about time schedule, but it cannot be done by anybody and in any situation. For some persons, this "alarm" could increase the pressure to be on time. Anyway, the good and experienced auditor can plan his activities pretty much realistically and in addition, he will develop skills of good time management.

Being late could happen very often, but having a situation when the auditor is finishing very fast in some areas will trigger an opinion with the company that the auditor does not know what he is doing. A balanced Audit Schedule with 10% spare time in each activity is highly required!

11.6.3 Honesty

I am not sure that honesty should be mentioned here... Not because it is not required, but because it is requirement for human lives in general.

In this section I would like to mention one dishonest auditor's company which was involved into one of the great scandals in US history. I am speaking about Arthur Andersen, one of the five world companies which were providing auditing, accountancy, and consultancy in different areas of economy and businesses. Arthur Andersen was involved in the Enron scandal (2001) when this company went into bankruptcy. Arthur Andersen helped Enron executives to hide billions of dollars in debts from failed deals and projects. Financial audit reports, which were provided by Arthur Andersen, were fabricated, making Enron a successful company, although they had problems. Later, when the scandal exploded and investigation started, Arthur Andersen was prosecuted for destroying documents relevant to investigation. This scandal actually closed Arthur Andersen and this company does not exist anymore.

The overall outcome of this scandal was changing the law for financial audits by increasing the accountability of auditors if they are biased or not impartial. However, Enron employees and shareholders lost billions of US dollars, mostly in pensions and stock prices.

I would not continue with honesty here, because I do believe that the Arthur Andersen story could provide enough data to show how the auditor should not be! And there is an axiom which must be presented here:

Axiom 24: An auditor may not under any circumstances produce an erroneous audit report!

People make mistakes and this is natural, but publishing an erroneous audit report is a crime and it must not be allowed. Doing this is a lack of self-respect by auditors. Under "erroneous report," I do not mean only the report with wrong findings, but also an incomplete report or wrong objective evidence inside. In Risky Industries, producing an erroneous audit report is a recipe for disaster!

Whatever the situation is, the Team Leader must check twice what is written in the Final Report! Even a wrong observation can change the future for the auditor, for the company audited, especially for Risky Industries and for humans, assets, and the environment.

11.6.4 Independence

Each auditor individually, even he is part of the Audit Team, must be independent in his conclusions. This independence comes from the responsibility: The auditor is responsible for the findings which must be based on Objective Evidence gathered by him!

The auditor may ask for the opinion of his Team Leader or of his colleagues, but however their opinions affect the decision he makes, it is his responsibility. He must not be involved in politics and economy: All that matters is the Objective Evidence which can prove compliance or non-compliance. His decisions must not waver or be influenced by anyone!

This independence is connected strongly with auditor's self-confidence. If he does not feel self-confident, he is losing independence in his decision-making. Self-confidence is fostered by knowledge and by skills gathered during education, training, and the audits conducted. By improving these things, auditor could achieve a respectable level of experience which will feed his self-confidence and it will provide necessary independence.

There is a normal process during the audit, where auditors communicate with each other or with Team Leader, but it is just with intention to exchange experience which they gathered from ongoing audit and to ask for advice from more experienced auditors. Anyway, auditors are responsible for their own independence, so they must use advice to build their own opinion about the situation. The final decision regarding findings and audits stays with the Team Leader, because he is producing the Final Report. However, the auditor is responsible for integrity of his report to the Team Leader.

11.6.5 Impartiality

If the auditor is connected with the company which is audited, it is called a "conflict of interest." The auditor must be honest with himself and he must not take part in any audit process where he or someone from his family has some connection with the audited company. It is wise and prudent to report it and to miss such an audit. This is required, because if something happens in the future, his report cannot be considered as impartial. I am not saying that auditor will have some "hidden agenda" during the audit. I am just saying that he must not be allowed to do audit in a company where such connections exist.

Auditors must be always impartial and this is not always easy to achieve.

There is another aspect of impartiality and it is built on the "personality" of the companies already established by the auditor. Having in mind that the audit is about Management Systems and humans, the auditor may not have prejudices about the company or the employees inside. Whatever the class of the company's "personality" is, there shall be no change in the auditor behavior or in the way how he conducts the audit. Each company must be given the same chance to prove compliance. He may use the company's "personality" only to build his expectations, but whatever the company is, the impartiality must be maintained during all processes and during each audit!

The important thing in maintaining impartiality is to stick to some principles. Accepting and following the axioms in this book as audit principles, I do believe will provide the necessary impartiality for each audit.

11.6.6 Communication

An auditor must be a good communicator! He must know how (verbally and in writing!) to present requirements, how and when to ask the right questions, and how to listen to the speakers in front of him. It is important to be good at communication, because the way the auditor communicates with others builds or loses the respect which he needs to do his job.

An important thing during communication is to prefer face-to-face communication. This is important, because words spoken by the counterpart person must be in agreement with his facial expressions and with body gestures. A good communicator appreciates face-to-face communication because this type of communication always provides more information than written communication.

In communication with auditees, the auditor must be assertive, but not arrogant. He must be convincing and his communication must be based on facts, especially when he is presenting findings. He must be reasonable with the facts and he has to provide additional data to support his findings. He may recall some events from the past to support his findings or to explain why the regulation is dealing with this issue. In general, good articulation during expression is important.

Communication is a two-way process and the auditor must not forget this. As he must build the skills to speak and write, he must build the skills to be active listener. Active listener does not mean only allowing auditees to say what they would like to say, but it means analyzing fast what has been said by them. This analyzing should be in the scope of the conducted audit. The auditee will try very hard to explain how the company achieved compliance and the auditor must hear and analyze in his mind all presented statements and facts. Active listener means also that he should ask any question if he is confused or not satisfied with the auditee statements and presented facts. An active listener will "feed" conversation to the auditee regarding the audit, with the intention to gather as much data as possible.

It is important for the auditor to make a difference between empty conversations and considerable explanations provided by the auditee. Sometimes the auditee may use empty conversations about topics which do not have anything to do with the audit, with the intention to turn out the auditor's attention from some problem(s).

There is an interesting relationship between the verbal and written communication. Verbal communication provides more information and can very easily point to the particular issues during the audit, but it cannot be used as Objective Evidence. Written communication (mostly through e-mails) cannot provide as much information, but what is written can be used as Objective Evidence. So, the auditor may use verbal communication to gather information, but only as tool as to what to investigate. In looking for compliance or non-compliance, he must deal with written documents. In general, the auditor must be able to use both of them (verbal and written communications)!

Axiom 25: An auditor may not under any circumstances allow himself to lose his temper with the auditee during the audits!

The most important thing is that auditor must be polite during each type of communication (verbal or written!). He must not allow himself to be offended by anything and he must not lose his temper over anything. He must not be offended because the audit is not about him, it is about the company, and the company will do anything to obtain good audit results. Sometimes this "do anything" from the companies, could be something hostile, immoral, or impolite, maybe even criminal! This is nothing strange: Anger and denial are natural reactions of humans when they receive bad news, as it can be a disappointing audit result. Whatever happens during the audit, the auditor must not be shocked and must not be offended. He must always be polite and he must always maintain a working calmness. Ultimately, he is in charge regarding the audit and he may not lose control!

11.6.7 Flexibility

An auditor must possess a considerable degree of flexibility. This flexibility does not apply to understanding the findings, but it does apply to the audit schedule and other arrangements.

Whatever the plan for audit (Audit Schedule!) is, it can fail many times! So, the particular level of flexibility must always be shown. It is good if the auditor in advance could plan possible situations which can go wrong, but it cannot be planned in general. Anyway, there is need to have a back-up schedule for each situation. A very important thing is to know if the audit is not going in accordance with the schedule. The worst situation is to prolong it for another day. There are also situations when particular operations cannot be executed every day in the company, so in such cases auditor must reschedule the audit. Most of the conflicting situations regarding the Audit Schedule can be handled during the Opening Meeting or even earlier, having in mind that that Audit Schedule will be send to the company before the audit.

The auditor must not always maintain flexibility. Sometimes he must intentionally not use it, and this is particularly important in the cases when his flexibility can be abused by auditees. A balanced approach in using flexibility is important!

Whatever happen, he must be flexible and adapt the situation in his favor!

11.6.8 Trustworthiness

The auditor is in the company during the audits and he must maintain a particular level of trustworthiness. Under no circumstances may the auditor share information about the company gathered during the audit with anyone from outside. Safety Management is not secret, but being in a company for few days, the auditor will have a chance to see new technologies (methods, methodologies, machines, systems, etc.) which are (maybe) a company secret and he must protect such an information. Actually, sharing particular sensitive information about the company or company's activities could be a matter of criminal investigation, so auditor must not put himself into such a situation.

Axiom 26: An auditor must not under any circumstances disclose any kind of information regarding the company which is the subject of an audits!

Auditors can have access to any kind of information. For example, auditing an SMS of an airline showed that their Management System does not fulfill all regulatory requirements. Nevertheless, the non-compliance issues could be minor or just observation in some cases, so if this information is shared with someone outside the Regulator (with today's mania for social networks!), it could be misinterpreted as "the Airline is not safe." This is completely wrong, but the public have no understanding of what it means and the panic about this statement could bring the airline into a very bad situation.

Sharing any information regarding the audited company can cause damage not only to company, but also to the Regulator, even to the industry. Simply, it is logical in such a situation for the airline to sue the Regulator and they will ask for compensation. I do not like to speak about how the Regulator in this situation will lose respect in the eyes of other companies.

Trustworthiness is an expression of the auditor's self-respect and he must maintain it on the highest level. If an auditor does not have respect for himself, he cannot expect respect from others!

11.6.9 Decisiveness

The auditor will be very often in a position to make a decision on many aspects of conducting the audit. So, he must be decisive and he must build a personality to provide decisions which are based only on the facts (Objective Evidence!) Decisions do not apply only to the data gathered, but they can apply also to the changing the schedule, how to proceed, or what to do if something pops up. With experience, he will become better in this activity, but it is important to maintain concentration and stay focused during the audit.

These things, when auditor must decide what to do, are a constituent part of each audit.

But there is one more aspect to this decisiveness: It is Human Factor! Deciding what to do, especially in situations where there is not enough data, is very stressful. It may affect the auditor's performance and he must be aware about it. So, in all such situations, it is important for the auditor to stick to the facts, to his instincts and especially to the audit principles!

11.6.10 Analytical and Understanding

Each of us would create different structure for any company and it is important to understand that if we do not like some structure, it does not mean that it is not serving the purpose. That is why it is important for the auditor to be analytical and open-minded to understand the company structure and the interconnections inside.

Having the capability to analyze the offered data is important. There are a lot of tools and methods which can be used for such an analysis. Some of them are simple, but most of them are complex and they need software to be used. The auditor during the Documentation Audit may use some of these tools and methods, but during On-Site Audits, he must use only his analytical mind, his knowledge, skills, and experience. People who are good in engineering and mathematics, in general, have good training for analytical thinking.

11.6.11 Persistence

The auditor must be persistent in his efforts to understand what is going on in particular company. It is not only important in the Risky Industries, but also in other industries. The job of the auditor is to do a good audit and he must stick to it. His persistence must be shown especially in the areas where the documentation provides poor understanding regarding the operation or the process. The auditor must not continue if he does not clarify all uncertainties during the audit.

But there must be limits to this persistence. The auditor must understand that he is a guest in the company during the On-Site Audit and he must abide to the company policies and company rules. He must be able to understand what is allowed and what is not allowed. Maybe sometimes the company will hide behind these policies and rules, but it does not mean that auditor may stop with his persistence. He must find another way to gather data which will help him understand if there is compliance or non-compliance. If there is no such way, then he must state that company must provide Objective Evidence that is not in accordance with their rules or their policies.

He must also be persistent during audits because he must document every finding and it is not an easy job. Very often, the detailed information may change the outcome of the audit and auditor must do everything to get it.

The auditor cannot always be successful in his persistence and there is nothing wrong with that. Sometimes he simply must know when to stop pushing, and then move on to another subject. But he will not forget. Later, his persistence must push him to try again. By using his experience and skills, he will try again to recover the potential argument left behind before.

11.6.12 Trust in the Instincts

Somewhere I read that "The instinct can be found in the nose of the mind." And I like this statement!

A good auditor with considerable experience could develop instincts which will help him. One sight, one look, one word, or one gesture from the auditee can provoke the auditor's instincts. And he must trust them!

The feeling regarding the possible situations is very important and it can be developed by dedicated auditor by time. In such situations, he must check all facts and ask for more if he has a feeling that it is important. Again: Do not use instincts for findings! Findings must be based on Objective Evidence! But the instincts could help you with investigations. They can give you a hint that something is wrong or something could happen in the future. And all these things should be investigated during the audits.

11.6.13 Commitment and Determination

"Every success starts with the strong commitment and determination!" – This is a statement which is used in sport to amplify the efforts of athletes for any type of achievement. I do believe this applies not only to sports, but also to life. Commitment and determination during audits are also very important!

The personal commitment of the auditor is very important in providing efficient and effective audit. This is like looking for a crime in detective stories: You are committed and determined to solve the criminal case and you are looking for any type of information which can explain to you how the crime was executed! The only difference is that in the case of audits, it does not matter who did the crime! The auditor must clarify only the "crime."

Aside from the area of conducting audits, personal commitment and determination is important also for the future development of auditor as expert in auditing. Knowledge, skills, and experience must be maintained and upgraded, because there is no progress for auditor if he is not committed and determined to self-improvement.

11.6.14 Professionalism

Professionalism is not a quality requested only for the auditor position. It is wide-spread requirement in industry, but with the auditor it has more "weight." Having in mind the nature of a job (dealing with audits in Risky Industries), the fact that lack of professionalism from the auditor could bring safety consequences cannot be neglected. The auditor must understand that he is protecting the public and environmental interest and his job always sends a message to the companies. If he shows a low level of professionalism, it will be understood by the companies that this level of professionalism is the standard in the State (or in industry!) and it will be maintained by them also.

An important part of this professionalism is professional skepticism. Auditors must be skeptical of every explanation provided by auditees during audits. Of course, that does not mean that they must check everything, but the things which do not provide enough reasonable understanding must be clarified. A good auditor will be satisfied when all anomalies, estimations, and uncertainties are clarified.

A balanced approach here is also of utmost importance: The professional skepticism must not be expressed in an offensive way!

11.6.15 Team Work

Audit (especially in Risky Industries!) is done by a Team of auditors. The reason that it is done by a Team is simply because it is a complex job and it will not be effective or efficient if it is done by individual. It means that each individual auditor is the member of the Audit Team, and as he has obligations to the auditee, he has also obligations to the Audit Team.

Whatever is going on during the audit, the auditor must help the Team members in fulfilling their job. Each piece of information, each event during the audit, or whatever situation which can influence the audit, are shared with the Audit Team, mostly in Audit Team meetings at the end of the day or during lunch.

All team members are different. They are different because each of the auditors are coming with different knowledge, skills, culture, backgrounds, and areas of expertise. This diversity provides different insights on the problems during audits, which fosters good solutions for the problems and increases the quality of outcomes of audits. Each auditor must know the strengths and weaknesses of all Team Members, so they can support each other or they can ask for advice from a particular member. This is especially very important for the Team Leader, as a person who guides the audit.

11.7 The Most Important Thing!

In previous sections, the qualities of the Safety Auditor were presented. There were explanations of how auditor should be and how he should not be. But in this section, I would like to emphasize one very important thing which must be maintained during audits by auditors...

It is presented by Axiom 27:*

Axiom 27: An auditor who is auditing the SMS of a company in a Risky Industry must not under any circumstances allow himself to offer or provide any comment, any advice, any suggestion, or any help to the companies which are the subject of a Safety Audit!

I (myself!) had a such an experience with some of the auditors and honestly, I did not feel well. One auditor INSISTED (intentionally written in capital letters!) for me to provide some document (for the Management System which was created by me!) and this document was not regulatory requirement. In addition, this document was totally needless for my company or for the Management System. He agreed that it was not a regulatory requirement, but he INSISTED that the document will help the company in the future. He was insisting too much, so eventually I produced this document (Never make your Regulator angry!) and I put it in the documentation. But I never received any information that someone in the company used this document. Anyway, the auditor was extremely happy that I accepted his suggestion!

For those who are involved in quality or financial audits, Axiom 27 could look very strange (and wrong!). This is because the auditors are (sometimes) engaged as consultants in these areas. In the scope of their consultancy, the first thing which they do is they conduct very a comprehensive audit of the company's Management System under consideration, known as Gap Analysis.† Later the results of this Gap Analysis audit are considered together with the company managers and consultant company usually offers a few solutions on how to minimize the findings in Gap Analysis. In such cases, one of the consultancy objectives is to see is there any opportunity for improvement, and it is normal if the consultant company offers suggestions.

---

* Actually, this axiom is valid for the overall regulatory safety oversight process, not only to audits!
† Gap Analysis is always done before implementation of any Management System in the companies. It will show what is done and what needs to be done!

Anyway, there is such a tendency even for the auditors of SMS in Risky Industries to provide such "help" for companies. Probably they feel themselves to be important and knowledgeable in providing suggestions or advice, but this is an elementary wrong.

There are very strong reasons for abiding by the Axiom 27 for Safety Regulators.

The first reason is: It is not the auditor's job to comment, suggest, advise, or help! The auditor's job is to do the audit with the intention of establishing the status of the company regarding compliance (or non-compliance) in the scope of required implementation and maintenance of a Safety Management System. Anything beyond this is not professional! If the auditor would like to help, he must use the power of the Regulator to produce a regulation which will stick to the requirements of Risky Industries, but it will take into consideration business interests of the companies (if they do not endanger safety!). A good way to help companies is to provide considerable training to them regarding the regulation. Providing a good understanding of regulation to companies will help them very much and also to the Regulator in the future.

The second reason is, the company has created this Management System taking into consideration her position on the market, her resources, and her plan for the future. The auditor is not familiar with all these things, so commenting (suggesting, advising, or helping) is not appropriate. Simply put: He does not have enough information on what was behind the company solution for such a system. Someone will say: I will ask all these things! Yes, the auditor may ask, but does the company like to discuss these things with an auditor? Think about that: It is not polite at all!

The third reason is: By commenting (suggesting, advising, or helping), the auditor is pressurizing the company to do it. At least, this is the company's understanding regarding any suggestion made by the auditor: Do this and you are compliant! Do not forget that the position of Regulator is understood differently by the companies than the Regulator understands himself. This is in line with my example from beginning of this paragraph: Never make your Regulator angry!

The fourth reason: Commenting, suggesting, advising, or helping one company auditor could put this company into a privileged position. You will say: But the auditor will help every company! My answer is: Does the auditor have time to do that (beside the fact that this is not his job!)? And the worst outcome: If the other companies find out that the Regulator provided some suggestions or advice to some company, they will think that the Regulator is ruining impartiality by this deed. The Regulator must take care "not to be mother for someone and stepmother for others"!

The fifth reason is the most important: The auditor, by commenting (suggesting, advising, and helping), undertakes the responsibility for the performance of the company in that area and this must not happen under any circumstances!

Imagine that the company's representative or manger (if there was a finding of non-compliance!) asks the auditor will they be compliant if they buy some system (equipment, machine, instrument, etc.)? The auditor says: Yes, you will be compliant.

They buy the system and they spend a lot of money, but the non-compliance is still present. Who is responsible for the failure of buying an expensive system and the non-compliance is still present? Or maybe a worse situation: An accident happened and this system did not prevent it! Or even the newly purchased system caused the incident or accident! Who will be responsible for the incident or accident?

Do not forget that in the case of any incidents and accidents, the Regulator is the subject of investigation also!

The auditor shall stay aside from all such activities: No comments, no suggestions, no advice, and no help! Even if asked, auditor must refuse any such activity. It does not mean that he would be rude when refuse, but it means that he would say: Sorry, I cannot comment (suggest, advise, or help) on this subject! Your company shall decide what to do and *maybe* you need to engage a consultant.

As you can notice, the word "maybe" is italicized, because it shows that the auditor must not recommend even such a thing! In such a situation, the rumors that the auditor has some "connection" with any consultant companies will spread immediately!

Having everything (aforementioned) in the mind, it is a matter of good behavior and professionalism for the auditor to stick to Axiom 27.

Final Words

I sincerely believe that you decided to read this book with intention! I hope that you have tried to improve your regulatory auditing skills, and by writing this book I tried to give the facts, reasons, and justifications why particular activities must be done.

I was (and still I am!) extremely unsatisfied with the way aviation auditors are doing their job!

Forgive me, but it is very frustrating to produce any type of management system and to meet (later) the auditor who does not have even a basic understanding what auditing is about. Before I decided to change something, I realized that even the courses which are offered in this area are mostly informational, and they are not pointing to the important things which I called Axioms and which are part of this book. I could not find a book regarding the audit on the internet or in bookshops which can satisfy my criteria, so I decide to write my own.

OK, I agree: Maybe this book is just a "try" by a desperate man to improve our world!

You may not agree with many things mentioned inside. You may even produce your own axioms which will suit your understanding of audit. My point is that, during past 15 years, I was on both sides (like an auditor and an auditee!) and I have experienced all these things on my "skin." I was thinking a lot about how to improve them, I was discussing it a lot with colleagues from industry and Regulators (auditors!), even with the auditors from Certification companies, and I hope that I have produced a book which will really help to improve the audits.

In my humble opinion, the Regulators (in the world) employ in auditing positions 40% staff with a good auditing background and 60% of the staff with good industry experience. The problem here is that the 40% with a good auditing background are mostly bureaucratic in their approach to audits and the 60% from those with good industry experience rely mostly on their knowledge and totally neglect the auditing skills. I hope that this book will help any of these Regulators and auditors to find a proper balance between auditing skills and industrial expertise.

I would like to emphasize here again a very important thing: I presented almost everything which I believe is important for audits, but what will be implemented by the readers depend on them. Following the axioms is a must, but again: You decide what will you follow!

Appendix: Axioms

Axiom No.	Text
1	Whichever type of safety is under consideration, the company has the primary responsibility to provide for the safety of their workers, customers, environment, premises, operations, activities, products and/or services offered!
2	The particular Management System is built by the company, and if there is regulatory requirement to have such a system, then the Management System must be approved by the Regulator through the audit or through accepting the Certificate of Compliance from any approved Certification Body!
3	There are no two equal Management Systems in the world! If they are the same, one of them is a COPY of the other and this will not work!
4	When the auditor is auditing the Management System, he is checking the existence, implementation, performance, and maintenance of System Procedures!
5	There are always many ways regulations can be satisfied by the company. The auditor must understand how the company is trying to comply with the regulatory requirements to be able to audit the company!
6	For an audit to be successful and to be able to improve a Management System, it must be impartial and independent! The impartiality and independence of audits are provided by auditors!
7	Establishing the "personality" of the company will help you with the strategy and tactics to organize and conduct the audit!
8	Safety Regulators and companies are not connected in a Master–Slave relationship! They are partners in providing safety!
9	The audit must be well-prepared and organized by Regulators (in partnership with the companies!) to be effective and efficient in regard to resources and time!
10	The auditor shall use the Check Lists during the audit! An audit without Check Lists is a "struggle for survival" with a very pessimistic forecast: The auditor does not know what to do and when and how to do it!
11	The good regulation should care only for effectiveness of the SMS! The Efficiency of the proposed solutions is concern only of the company, not of the Regulator!
12	The Approval Audit is used to check if 100% compliance with the regulation is achieved by the company's SMS!
13	Whatever you think about the audit process, an audit can be defined as a quest for compliance! Whatever you think about the audit process, it is not a quest for non-compliance!
14	The auditor must not under any circumstances express or take into consideration his own opinion about the things (tasks, procedures, activities, operations, etc.) in the company which is the subject of the audit! All his statements or comments must be based on objective evidence!

Axiom No.	Text
15	Expertise in a particular area is not the only requirement for the Team Members of Audit Teams. Equal importance is given to the requirement to provide proper auditor training for each Team Member!
16	If the Audit Team is not satisfied with all answers during the Documentation Audit, the company's SMS manual must not be approved!
17	If the SMS manual is not approved, there is no reason to proceed with the On-Site Audit!
18	Due to the complexity of the job to be done during the On-Site Audit, the auditor uses Sampling to check processes (procedures, operations, records, etc.) in the company!
19	The auditor is the person who is choosing from where, how many, and which samples will be taken! He must not accept the samples offered by company's representative!
20	A Safety Policy explains the Strategy of how the company is dealing with safety! The tactics can be found in the rest of SMS Manual!
21	How the company is dealing with the changes in their operations is a good indicator of how the company is dealing with the implemented SMS in total!
22	Establishing the "Personality" of the company by the auditor is needed to provide information how the auditor will behave and what questions and when to ask them during On-Site Audits! The "Personality" of the company may not affect the findings!
23	A person without proper audit training may not do auditing!
24	An auditor may not under any circumstances produce an erroneous audit report!
25	An auditor may not under any circumstances allow himself to lose his temper with the auditee during the audit!
26	An auditor must not under any circumstances disclose any kind of information regarding the company which is the subject of an audit!
27	An auditor who is auditing the SMS of a company in a Risky Industry must not under any circumstances allow himself to offer or provide any comment, any advice, any suggestion, or any help to the companies which are the subject of a Safety Audit!

Index

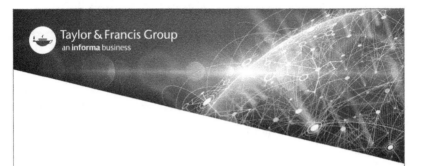